SPOTSYLVANIA COUNTY [VIRGINIA] ROAD ORDERS

1722-1734

Virginia Genealogical Society
Richmond, Virginia

Published With Permission from the

Virginia Transportation Research Council
(A Cooperative Organization Sponsored Jointly by the Virginia
Department of Transportation and
the University of Virginia)

HERITAGE BOOKS
2008

HERITAGE BOOKS
AN IMPRINT OF HERITAGE BOOKS, INC.

Books, CDs, and more—Worldwide

For our listing of thousands of titles see our website
at
www.HeritageBooks.com

Published 2008 by
HERITAGE BOOKS, INC.
Publishing Division
100 Railroad Avenue #104
Westminster, Maryland 21157

Copyright © 1985, 2004 Virginia Genealogical Society

All rights reserved. No part of this book may be reproduced or transmitted in any form or by any means, electronic or mechanical, including photocopying, recording or by any information storage and retrieval system without written permission from the author, except for the inclusion of brief quotations in a review.

International Standard Book Number: 978-0-7884-3671-0

SPOTSYLVANIA COUNTY ROAD ORDERS 1722-1734

By

Nathaniel Mason Pawlett
Faculty Research Historian

Virginia Highway & Transportation Research Council
(A Cooperative Organization Sponsored Jointly by the
Virginia Department of Highways & Transportation and
the University of Virginia)

Charlottesville, Virginia

January 1985
Revised April 2004
VHTRC 85-R17

Historic Roads of Virginia

Louisa County Road Orders 1742-1748, by Nathaniel Mason Pawlett. 57 pages, indexed, map.

Goochland County Road Orders 1728-1744, by Nathaniel Mason Pawlett. 120 pages, indexed, map.

Albemarle County Road Orders 1744-1748, by Nathaniel Mason Pawlett. 52 pages, indexed, map.

The Route of the Three Notch'd Road, by Nathaniel Mason Pawlett and Howard Newlon, Jr. 26 pages, illustrated, 2 maps.

An Index to Roads in the Albemarle County Surveyors Books 1744-1853, by Nathaniel Mason Pawlett. 10 pages, map.

Brief History of the Staunton and James River Turnpike, by Douglas Young. 22 pages, illustrated, map.

Albemarle County Road Orders 1783-1816, by Nathaniel Mason Pawlett. 421 pages, indexed.

A Brief History of the Roads of Virginia 1607-1840, by Nathaniel Mason Pawlett. 41 pages, 3 maps.

A Guide to the Preparation of County Road Histories, by Nathaniel Mason Pawlett. 26 pages, 2 maps.

Early Road Location: Key to Discovering Historic Resources? by Nathaniel Mason Pawlett and E. Edward Lay. 47 pages, illustrated, 3 maps.

Albemarle County Roads 1725-1816, by Nathaniel Mason Pawlett. 98 pages, illustrated, 5 maps.

Backsights: A Bibliography, by Nathaniel Mason Pawlett. 29 pages.

Orange County Road Orders 1734-1749, by Ann Brush Miller. 323 pages, indexed, map.

A Note on the Methods, Editing and Dating System

By

Nathaniel Mason Pawlett
Faculty Research Historian

The road and bridge orders contained in the order books of Spotsylvania County are the primary source of information for the study of its roads. Those from 1722 to 1734 were extracted, indexed and published by the Virginia Highway & Transportation Research Council. All of the county court order books were in manuscript, sometimes so damaged and faded as to be almost indecipherable. Rendered in the rather ornate copperplate script of the time, the phonetic spellings of this period often served to further complicate matters for the researcher and recorder.

The amount of material to be handled, as well as the number of people involved, rendered virtually useless the idea of longhand transcription. This same problem had been faced at the beginning of the Albemarle road study some years before, although the difficult nature of the Louisa records had initially forced the author to resort to longhand transcription for them. Shortly, however, in order to facilitate the work, a system was devised to render these road orders literatim into a small hand-held tape recorder and reproduce for transcription all their eighteenth-century idiosyncrasies. This recording system might be set out in the following rather general rules:

1. Capitalisation to be so stated. Viz: "cap Wadlow cap Cuthbert cap Twaddle" in the case of the name Wadlow Cuthbert Twaddle and any other words in the road order which are capitalised.

2. Names with variant or phonetic spellings to be spelled out. Do not assume that the common Virginian spelling is the only one.

3. No periods unless so stated.

4. Periods, commas, colons, semicolons, etc., to be stated. All symbols to be described as nearly-as-possible.

5. Date and pagination in order book, vestry book, etc., to be stated. Pre-1752 dates to be designated O.S. for Old Style.

6. Each new citation to be so stated.

7. No paragraphing unless so stated.

8. Tape reels to be marked sequentially as completed, with county, name of record, book number(s), and approximate date covered.

Following this, the orders were put into typescript by a secretary who had to unlearn many of the modern rules of spelling and then learn to render the orders exactly as dictated. Then the transcripts were compared to the tapes or to the original manuscript sources and corrected. Placed in their final form and indexed, the road orders were ready for publication and distribution.

With these road orders and the ones for Orange available in an indexed and cross-referenced published form, it will be possible to produce chronological chains of road orders illustrating the development of many of the early roads of a vast area from the threshold of settlement up through the middle of the eighteenth century. Immediate corroboration for these chains of early road orders will usually be provided by other evidence such as deeds, plats and the Confederate Engineers maps. Often, in fact, the principal roads will be found to survive in place under their early names.

With regard to the general editorial principles of the project, it has been our perception over the years as the road orders of Louisa, Hanover, Goochland and Albemarle have been examined and recorded that road orders themselves are really a variety of "notes", often cryptic, incomplete or based on assumptions concerning the level of knowledge of the reader. As such, any further abstracting or compression of them would tend to produce "notes" taken from "notes", making them even less comprehensible. The tendency has therefore been in the direction of restraint in editing, leaving any conclusions with regard to meaning up to the individual reader or researcher using these publications. In pursuing this course we have attempted to present the reader with a typescript text which is as near a type facsimile of the manuscript itself as we can come.

Our objective is to produce a text that conveys as near the precise form of the original as we can, reproducing all the peculiarities of the eighteenth century orthography. While some compromises have had to be made due to the keyboard of the modern typewriter, this was really not that difficult a task. Most of their symbols can be accommodated by modern typography, and most abbreviations are fairly clear as to meaning.

Punctuation may appear misleading at times, with unnecessary commas or commas placed where periods should be located; appropriate terminal punctuation is often missing or else takes the form of a symbol such as a long dash, etc. The original capitalisation has been retained insofar as it was possible to determine from the original manuscript whether capitals were intended. No capitals have been inserted in place of those originally omitted. The original spelling and syntax have been retained throughout, even including the obvious error in various places, such as repetitions of words and simple clerical errors. Ampersands have been retained throughout to include such forms as "&c.^{a}" for "etc." Superscript letters have also been retained where used in y^e., y^t., s^d. The thorn symbol (y), pronounced as "th," has been retained in the aforesaid "y^e.", pronounced "the", and "y^t." (that), along with the tailed p which the limitations of the modern typewriter have forced us to render as a capital "p". This should be taken to mean either "per" (by), "pre" or "pro" (sometimes "par" as in "Pish" for parish) as the context of the order may demand. For damaged and missing portions of the manuscript we have used square brackets to denote the [missing], [torn] or [illegible] portions. Due to the large number of ancient forms of spelling, grammar and syntax it has been deemed impracticable to insert the form [sic] after each one to indicate a literal rendering. Therefore, the reader must assume that apparent errors are merely the result of our literal transcription of the road orders, barring the introduction of typographical errors of course. If in any case this appears to present insuperable problems, resort should be made to the original records available for examination at Spotsylvania Court House.

As to dating, most historians and genealogists who have worked with early Virginian records will be aware of the English dating system in use down to 1752. Although there was an eleven-day difference from our calendar in the day of the month, the principal difference lay in the fact that the beginning of the year was dated from March 25 rather than January 1, as was the case from 1752 onward to the present. Thus January, February and March (to the 25th) were the last three months in a given year and the new year came in only on March 25.

Early Virginian records usually follow this practice, though in some cases dates during these three months will be shown in the form 1732/3, showing both the English date and that in use on the Continent, where the year began January 1. For researchers using material with dates in the English style it is important to remember that under this system (for instance) a man might die in January 1734 yet convey property or serve in public office in June 1734, since June came before January in a given year under this system.

SPOTSYLVANIA COUNTY ROAD ORDERS 1722-1734

by

Nathaniel Mason Pawlett
Faculty Research Historian

INTRODUCTION

The roads are under the government of the county courts, subject to be controuled by the general court. They order new roads to be opened whenever they think them necessary. The inhabitants of the county are by them laid off into precincts, to each of which they allot a convenient portion of the public roads to be kept in repair. Such bridges as may be built without the assistance of artificers, they are to build. If the stream be such as to require a bridge of regular workmanship, the court employs workmen to build it, at the expense of the whole county. If it be too great for the county, application is made to the general assembly, who authorize individuals to build it, and to take a fixed toll from all passengers, or give sanction to such other proposition as to them appears reasonable.

Thomas Jefferson,
Notes on the State of Virginia, 1781.

The establishment and maintenance of public roads was one of the most important functions of the County Court during the colonial period in Virginia. Each road was opened and maintained by an Overseer of Highways appointed by the Gentlemen Justices yearly. He was usually assigned all the "Labouring Male Titheables" living on or near the road for this purpose. These individuals then furnished all their own tools, wagons, and teams and were required to labour for six days each year on the roads.

Major projects, such as bridges over rivers, demanding considerable expenditures were executed by Commissioners appointed by the Court to select the site and to contract with workmen for the construction. Where bridges connected two counties, a commission was appointed by each and they cooperated in executing the work.

At its inception Spotsylvania County comprised a large part of the Piedmont frontier east of the Blue Ridge, running to the Shenandoah River in the Valley. Spotsylvania was described by the 1720 act providing for its creation as bordering "upon Snow Creek up to the Mill, thence by a south-west line to the river North-Anna, thence up the said river as far as convenient, and thence by a line to be run over the high mountains to the river on the north-west side thereof, so as to include the northern passage thro' the said mountains, thence down the said river until it comes against the head of Rappahannock, thence by a line to the head of Rappahannock river; and down that river to the mouth of Snow Creek…"

Formed from the upper ends of the counties of Essex, King William and King and Queen, the eastern part of Spotsylvania already had undergone some settlement and rudimentary road building. West of the mountains in the Valley, settlement would shortly get under way. The years 1722-1734 encompass the period when Spotsylvania was a giant parent county stretching to the middle of the Valley. By the separation of Orange in 1734 it was reduced to its present size. But for the twelve years from 1722 to 1734 it contained within its bounds the present Piedmont counties of Orange, Culpeper, Madison, Rappahannock and Greene, as well as the Valley counties of Rockingham, Page and Warren.

The road orders contained in this volume cover the period from Spotsylvania's creation to the creation of Orange County in 1734. As such, they are the principal extant evidence concerning the early road development of a major proportion of the northern Virginia Piedmont.

THE DEVELOPMENT OF SPOTSYLVANIA COUNTY

Note: As originally published in paper format, this volume included maps showing the evolution of the county. Maps are not included in the revised/electronic version due to legibility and file size considerations. Instead, a verbal description is provided.

Prior to 1720, the area that would become Spotsylvania County was part of the far western reaches of Essex County, King & Queen County, and King William County. As formed in 1720, Spotsylvania County included the present counties of Spotsylvania, Orange, Greene, Culpeper, Madison, and Rappahannock east of the Blue Ridge Mountains, and a portion of the Shenandoah Valley as far west as the Shenandoah River.

From 1720 to 1728, Spotsylvania County's southeastern border adjoined the remaining territories of its parent counties. In 1728, the western portions of Essex, King & Queen, and King William counties were further divided to form Caroline County, bordering Spotsylvania to the southeast.

Spotsylvania County reached its present form in 1734, when all of the territory of Spotsylvania County west of Wilderness Run, became the newly-created Orange County. A giant county at its formation, Orange included the present-day counties of Orange, Greene, Culpeper, Madison, and Rappahannock east of the Blue Ridge Mountains, as well as the territory west of the mountains extending, at least nominally, to the Mississippi River.

Will Book A

7 August 1722 O.S., Page 1
Lawrance Taliaferro petitioning this Courte for the Road to be cleared from the little mountains to the wilderness bridge, In this County, is Granted, and William Eddings is appointed overseer of the same, and all ye Inhabitants above the Mountain run, are ordered under him to help clear the same --

7 August 1722 O.S., Page 1
Thomas Carr Junr: & c pettitioning for a Road to be cleared from the mouth of East north East to Germanna as is already Marked and laid ut, is granted & John Hawkings is appointed overseer of the same, and all ye Inhabitants of that part of ye county that lives between Mattapony and North Anna are ordered under him to help clear the same from East North East to ye road of Mattapony [indistinct] --

4 September 1722 O.S., Page 3
Thomas Watts is appointed Overseer of the road from the falls to the wilderness bridge in the room of John Taliaferro Gent^n: & the same tithables that Served on ye Said road, are ordered to serve under the said Watts to clear the same --

4 September 1722 O.S., Page 3
James Canny is appointed overseer of the lower part of the new road, and the same tithables as was on the Said road before are ordered to serve under the said Canny to clear the same --

4 September 1722 O.S., Page 3
Philemon Cavenough is appointed Overseer on the road from Massaponax to the falls, and the same tithables as were on the said road before, are ordered to Serve under the said Cavnough to clear the same --

4 September 1722 O.S., Page 3
Philemon Cavenough is appointed Overseer of the upper part of the new road and the same tithables as were on the said road before, are ordered to serve under the Said Cavenough to clear the same --

4 September 1722 O.S., Page 4
William Bartlett petitiond that a road may be cleared from the bridge on the river Noy below John Lewis Esq^{r}s quarter to his Hon^{bl} ye Governors quarter called the bridge Quarter. ye S^d petitioned was rejected.

5 September 1722 O.S., Page 4
Edward ffranklyn petitiond for a Road to be cleared from the head of Greens branch crossing the river ney to the Germana road: which was granted and ordered that the said Edward ffranklyn be Overseer thereof --

5 September 1722 O.S., Page 4
Larkin Chew Gent^n: petitiond for a Road to be cleared from John Robinsons Esqr quarter over the river Po by the Said Chews to Massaponax roleing house known by the name of James Kenny which was granted and Capt: Larkin Chew is ordered & appointed Overseer [page edge eroded]

2 October 1722 O.S., Page 9
Upon Petition of William Bartlet & c for a Road to be cleared from the Hon$^{\&\ ble}$: Alexander Spotswoods Quarter called Bridge Quarter, down the Ridge between the River Po and Ny & over the Same by Lewis bridge to the Road cleared P King & Queen Inhabitants. is granted by the Court, and the Said William Bartlett is appointed Overseer thereofe

2 October 1722 O.S., Page 9
Upon Petition of Augustine Smith gent^n: for a Road to be cleared from Germana along a Ridge to the Mountain Run, So a Cross. ye Said Run & to goe between Coll^o: Carters & Skreens Quarter is granted by the Court & Jacob Wall is appointed overseer thereof ---

4 December 1722 O.S., Page 13
On Petition of Jacob Wall Overseer of ye road from Germana along the ridge to the mountain Run so a Cross the said run between Coll^o: Caterss & Mr Skreens Quarter petitioning this Court & complaining against Several persons hereafter mentiond ffor not appearing and helping him as overseer appointed by this Court, to Clear the Said road upon his

warning & summonds, the Court affter examining the said persons, ordered as followeth That

    William Aginfeild be fined for absenting himself two days –
    William Bolen for ffor three days
    John Bolen for five days
    David Bolen for five days
    Abram. ffeild for five days

and that thay pay for each days absence and contempt as the Law directs with Costs as the Law directs with Cost alias: John Hows and William Thomas makeing it appear that they were disabled P sickness were excused for their absence & complaint of ye Said Jacob Wall Overseer

2 April 1723 O.S., Page 18
On Petition of William Bartlett Surveyor of the bridge Quarter road to King and Queen County Line Setting forth that the said road being about twenty miles in length was too much trouble for one Overseer to cleare it, its therefore considered, & ordered, that Robert King be appointed surveyor of that part of the said road, Viz^t: from King & Queen County road to the lower branch of Greens branch & that all the tithables belowe Mr Edwards Quarter are to assist him to clear the same

And that Bartlett be continued Surveyor of that part of the said road from the said Greens branch to Coll^o: Corbins roleing road & all the tithables from thence to Cow Land are to Assist him clear the same –

and that William Watters be appointed Surveyor ffrom Cow Land to the bridge Quarter that comes into Germana road and all the tithables between ye Said places are to assist him clear the same --

2 April 1723 O.S., Page 18
On Petition of Edward franklin Survevor of the road from the head of Greens branch to Germana old road: for people to assist him clear ye Same Ordered that all the tithables belonging to the middle precints doe assist him clear the same --

7 May 1723 O.S., Page 27
On Petition of sundry inhabitants that lives on the North Anna River In this County that a new Overseer of the road from the devideing line to Rich Neck may be appointed, Ordered that Thomas Gambal be Overseer of the said Road, and all the tithables which formerly belonged to the s^d road, & those which are since Come in this County that lives on the said River are appointed to help him clear ye Same

7 May 1723 O.S., Page 28
On Motion of Thomas Chew Gent^n: all the tithables that are below Edward ffranklins are ordered to help Robert King Clear his road, instead of the tithables y^t: are mentioned P this Courts order ye 2^d: day of April last --

2 July 1723 O.S., Page 36
John Purvis is appointed Overseer on that part of the road ffrom Snow Creek to Massaponax In this County and all the tithables that are between the said places are ordered to help him clear the same --

3 September 1723 O.S., Page 44
Richard Sharp is appointed Surveyor on the road from the wilderness bridge to the pine Stake in the room of William Eddins, who is discharged of the same --

3 September 1723 O.S., Page 44
John Hawkings is discharged of being Surveyor of the Road from the mouth of East North East to the head of Mattapony River that goes toward Germana and John Kingbrow is appointed Surveyor in his room and all the tithables of the County that lives between North Anna and Mattapony are Ordered to help him cleare the same

3 December 1723 O.S., Page 57
Alexander Carr is appointed Surveyor of the road from the Stones marked AL belowe the Wilderness bridge to Germana, and it is ordered that Thomas Watts and the gang under him meet and assist the said carr & his gang to mend the wilderness bridge --

7 April 1724 O.S., Page 66
On Petition of John Kingbrow overseer of the Road from East North East to the head of Mattapany river towards Germana against Severall people, Viz^t: W^m: Crany, Thomas Mauldin, Edward ffranklin overseer of two negroes belonging to m^r Booker, William Lindsey, John Bush, Robert King, John ffoster, Samuell Moore, m^r Harry Beverly, Thomas Warrin William Warrin, Samuell Ham, Barnett Pain, Lawrence ffrenklin, Thomas Appason Thomas Hubbard, M^r: Goodlow, Mark Wheeler, and John Pigg, for not appearing to help clear the said road According to Summonds & c as the petition and List setts forth--

Ordered that the said Severall people be Summond to the next court to Shew cause (if any) why thay Shall not be fined as the Law directs --

2 June 1724 O.S., Page 73
On petition of Edward ffranklin to be released from being overseer of the road from the head of greens branch crossing the river Ney to the Germana road, was granted, and William Richardson is appointed Overseer in his room--

7 July 1724 O.S., Page 80
On Petition of Thomas Watts he is discharged from being Overseer of the road, that goes from the falls to the wilderness bridge --

7 July 1724 O.S., Page 80
Ordered that Robert Slaughter Junr: be overseer of the road, from the Stones, to the road that goes to William Eddins below the orchard of white oakes, and that Lewis Elzeys people be added to his gang to help clear the same --

7 July 1724 O.S., Page 80
Ordered that John Mulkey be overseer of that [missing] of the road that goes to William Eddins below the orchard of white Oaks, from thence down to the fall landing

7 July 1724 O.S., Page 81
On Petition of Phillemon Cavenaugh he is discharged from being Overseer of the new road that goes from the German road to the division, and Henry Martin is appointed overseer in his room, & ordered that the Inhabitants of Massaponax and George Procters doe assist him to help cleare the same --

7 July 1724 O.S., Page 81
On Petition of Goodrich Lightfoot Gent^n: & c for a Road from Coll^o: Alex^r: spotswood litle mill to Thomas Stantons, was granted, and ordered that George Lightfoot, m^r George Wheatlys, m^r Thomas Stantons, m^r Henry ffeilds & m^r William Neal tithables doe help clear ye Same.

7 July 1724 O.S., Page 83
On Petition of John Kimbrow against Sundry people for not appearing according to summonds to help clear his road the court on consideration

of the great remoteness and distance thay have to come to ye Said road, thought it at this time unnessessary, therefore ordered that the said Petition be dismist --

7 July 1724 O.S., Page 83
Ordered that John Kimbrow be continued Overseer of the said road from the mouth of east north east to the head of Mattapany river towards Germana and that the inhabitants of Pamunky river, and of east north east do help him clear the same --

8 July 1724 O.S., Page 86
On Petition of Cat^t: Larkin Chew he is discharged from being overseer of the road from John Robinsons Esqr: Quarter over the river Po, by the said Capt: Larkin Chews to Massapanax Roleing house, Known by the name of James Kennys, and Samuell Moore is appointed in his room--

8 July 1724 O.S., Page 86
William Bartlett overseer is ordered to clear his road up to William Watters and Coll^o: Corbins upper road --

Spotsylvania County Order Book, 1724-1730

6 October 1724 O.S., Page 16
On petition of John Purvis he is Discharged from being Surveyor of the Road from Snow Creek to Massaponax, John Cammell is therefore appointed in his room--

6 October 1724 O.S., Page 16
On petition of m^r: Ambrose Madison for a road from the Pine Stake to his plantation & Capt: John Taliaferros Quarter is granted, and Ordered that Thomas Phillips be Overseer thereof, and all the Male Tithables that lives above the said Capt: John Taliaferros quarter do help him clear the same --

6 October 1724 O.S., Page 17
On Petition of William Russell in behalf of himself & others the Inhabitants of this county to have a Road from Franklyns road to the new chappell now a building and so from thence to east north east bridge is granted, and it is therefore ordered that William Bartlett & William Brandegun doe veiw and marke that part of it from Franklyn road to the new Chappell & that John Gambrell & ffrancis Arnold doe veiw & mark out that part of it from the new church to east north east bridge the most convenients & least prejudiciall to any proprietor of Land and that they make returns of their proceedings to the next court --

3 November 1724 O.S., Page 27
On Petition of William Beverly Gent^n: in behalfe of him self & others for a road to be cleared from the end of the road whereof Jacob Wall is overseer down to the Island ford & to Beverly's quarter, ordered that Robert Green, Abram Feild, and Francis Kirkley or any two of them do lay out and make the most convenient way for a road, haveing regard to lay it out the least prejuducial to any propriators of Land and make returns of their proceedings to the next Court --

3 November 1724 O.S., Page 27
On the returne of William Bartlett and William Brandegon of the order of Court for the laying out the road from the new church on the river Ta to Franklyns road on the head of greens branch the same is approved P the court and ordered that Edward Franklyn be overseer thereof, and that the following tithables do help clear the same, Viz: Joseph Brocks, Major Benj^a: Robinson, William Branegons, Capt: Richard Booker, John Snell John Taliaferro, & William Hutcheson, plantations and all the male tithables thereunto belonging --

3 November 1724 O.S., Page 27
Maj: Augustine Smith is appointed & made overseer of the road from Germana along the ridge to the mountain run to cross the said run between Coll°: Carters & Skreins Quarter: in the room of Jacob Wall --

4 November 1724 O.S., Page 30
Ordered that Thomas Gambrell be discharged from being overseer of the road from the county line to Rich neck, and that Edwin Hickman Gent^n: do serve in his room and that the following Male tithables on Severall plantations do Serve & help him cleare the Same Vizt: m^r Augustine Moores lower quarter, Capt: Tho: Carrs quarter, Francis Arnold, W^m: Lobs, John Trusty, Edwin Hickman Gent^n: Thomas Gambrel, John Gambrell, Capt: William Smiths quarter, and Edward Downes --

4 November 1724 O.S., Page 30
John Kimbrow is appointed & continued overseer of the road from Rich Neck to the extent up wards to ye head of Mattapany river and Ordered that the following male tithables on the severall plantations do serve and help him cleare the same Viz^t : John Kimbrows, John Lawlys, m^r Augustine Moore, upper quarters Thomas Sartain, John Hollidays, John Keys, Robert Turner, Capt: Jerimiah Clowder William Craney, John Bush & Capt: Hankings Quarter--

2 March 1724/5 O.S., Page 38
On Petition of William Beverly Gent^n: in behalfe of himselfe & others for a road to be laid out, and cleared from Maj^r: Augustine Smiths road the most convenients way to the upper inhabitants is granted, and Robert Green, Francis Kirkeley, & Abram ffeild, or any two of them are ordered to layout the same & make returne of their proceedings to the next court --

2 March 1724/5 O.S., Page 39
On petition of Harry Beverly Gentl^m: that ye road may be cleared from Capt: Larkin Chews bridge up to Germana road, the same is granted & ordered that Samuele Moore with his gang do clear the same --

2 March 1724/5 O.S., Page 39
Mr John Finleson is appointed Overseer of the road from the stones below wilderness bridge to Germana Ferry in the room of Alexand Carr (who is removed out of this county) & therefore ordered that he doe clear the same, with the same gang as was under the said Carr --

3 March 1724/5 O.S., Page 41
On Petition of Henry Goodloe Gent[n]: in behalf of himselfe & others for a road to be laid out & cleared from his hous to the new Church that is built on the river ta, alias midle river is granted and ordered that William Warren, Thomas Warren, and mark Wheeler, or any two of them do layout a road to the said Church the most convenients for the petitioners & least prejudiciall to the proprietors of the lands, through which the said road will run, and make return of their proceedings to the next Court --

6 April 1725 O.S., Page 44
M[r] Henry Goodloe makeing returne of the order of last court for a road from his house to the Church on the River Ta, which was not quite full enough, Ordered that the road according as it is marked & laid out & so through the lands of Barnett Bain and Edward Pigg to the most convenienst way to Collo: John Robinsons rolling road, be the road and that Lawrence Franklyn be surveyor of the same, and that the male tithables of Henry Goodloe Gent[n]:, Mark Wheeler George Pemberton, Samuell Hamm, Mr William Stanards Quarter, Thomas Warrens, John Askew William Rice, Barnett Bain, Mr Nathaniell Sanders Quarter & John Bains plantations do help him clear the same –

4 May 1725 O.S., Page 47
On Petition of m[r] Harry Beverly for a road from his house to the Church on the River Ta (alias midle river), it is granted in order that John Foster be overseer of the same, and that M[r] Harry Beverlys, M[r] Richard Buckners quarter, Cap[t]: Larking Chews, Cap[t]: Thomas Chews, Cap[t]: Joseph Smiths, Abram Brown Robert King Alexander Cleveland, male working tithables do help the said Foster clear the same the most convenients way --

1 June 1725 O.S., Page 53
The veiwers haveing made their report about the road y[t]: William Beverly Gent: petitiond for, its therefore ordered that the road as the viewers have so laid out, be a road for the said petitioners --

6 July 1725 O.S., Page 63
On Petition of Michell Cock and Henry Snyder & other of y[e] Germans for to have leave to lay of, and clear a road from the ferry at Germana to Smiths Island up the river rapadan, the same is granted & ordered that they have liberty to lay of, and cleare the same --

5 October 1725 O.S., Page 76
On Petition of Lawrence Franklyn he is discharged from being overseer of ye Road from m^r Henry Goodloes to the church on ye river Tay and Samuell ham is appointed to serve in his room--

5 October 1725 O.S., Page 76
On Petition Coll^o: Gawen Corbin for to have liberty to cleare a Road from his Stone hill quarter by William Eddins to the german road, is granted.

6 October 1725 O.S., Page 78
Ordered that a road from the ferry at Germana, be cleared up the value the most convenients way below the first bridge to ye German road and that m^r John Finleson with his gang do clear the same --

6 October 1725 O.S., Page 80
On Petition of Richard Sharp it is ordered that the hands should be male tithables belonging to ye Honble: Alexander Spotswood att ye Chestnutt quarter be added to his gang to help clear the mountain run road --

2 November 1725 O.S., Page 83
On petition of Henry Martin over Seer of the Mausauponax road, that his & Edward Prices, Charles Stuart and James Sparkes, male tithables may be added to his gang to assist him to clear and grub the same, is granted, therefore ordered that they doe accordingly help & assist the said Martin clear the same --

2 November 1725 O.S., Page 84
Wee likewise present the Surveyor of the Wilderness run bridge, in Spotsylvania County, for not keeping it in repair according to Law--

...

It is therfore ordered that the said Overseer John Finleson be summoned to the next court to answer the said presentment of the grand jury for not keeping the wilderness run bridge in repair according to law--

3 November 1725 O.S., Page 86
Robert Green in behalf of him Selfe & Severall others petitioning for a bridge to be made over the Mountain run in the fork of Rapahanack river att the county charge, was Rejected

3 November 1725 O.S., Page 87
On Petition of William Beverly Gent[n]: Peter Russell is appointed Overseer of the new road from the Mountain run in the fork to the upper inhabitants and Ordered that all the Inhabitants above y[e] Mountain run do help him cleare & grub the same, and that both gangs do meet to help make & build the bridge over the said Mountain run--

3 November 1725 O.S., Page 88
On Petition of John Taliaferro Gent[n]: for leave to turn the road over massauaponax for the better to make a convenient bridge over the said run is granted & the S[d] John Taliaferro Gent[n]: is appointed Overseer of the same, & ordered that both gangs do meet to help make the bridge, and that the people below deep run do clear the road from the bridge upwards --

3 November 1725 O.S., Page 88
It is ordered that all the people above Massauponax do Serve under James Roy (and them only) on the new road to help him clear the same --

3 November 1725 O.S., Page 88
On petition of John Quarles, Charles Stephans is appointed overseer in the room of William Richeson (& that the S[d] Richardson be discharged of the Same) of the rowling road from Mattapony main road, to the massauponax road, and that the said Stephens have liberty to turn the road the most convenients way for a bridge into Massauponax new road, and that the former order of the old German road be made voyd --

2 February 1725/6 O.S., Page 100
Henry Goodloe Gent[n]: makeing Returne of the order of Court (Impowering John Taliaferro Edwin Hickman Gent[n]: & himself to agree with any person to build a bridge over the River Po & e[a]: and agreement made with cap[t]: Larkin Chew, Ordered that y[e] said agreement be Recorded & that y[e] Said Cap[t]: Larkin Chew do give bond & Securety on Demand whensoever the Court requires the Same for the performance of ye Said Agreement --

In Persuance of tha within order we the Subscribers Meett at Cap[t]: Larkin Chew & Veiw[d]: The Roads & River and find the most Convenient place, to be The Long Point above Abraham Brown's to build a bridge over The River Po & Is The Convenient's way for the Inhabitants of the South Side thereof and we have agreed with Cap[t]: Larkin Chew to build a bridge according to Law att eh Said place, and to give him for building y[e] said bridge & keeping it nine years in good Repair, (and Deliver it passable and in Repair att the end of nine years from y[e] finishing the Said bridge) and In Consideration of ye said agreement wee do agree to have Levyed att the Layng of the gext Levy In this County The Just Sum of Eight Thousand pounds of Tob°: with Cask to be paid within this county, If the Said Bridge be buildt and be passable by the Last Day of May next, Wittness our hands: Novem: The 9th: 1725.

               John Taliaferro
               Edwin Hickman
               Henry Goodloe

2 February 1725/6 O.S., Page 101
Ordered that the new Road from ye Mountains which Peter Russell was made overseer of, and the Road that goes from Germana to ye Mountains as Maj^r: Augustine Smith was appointed Overseer of, be made into one and that the said Peter Russell & Maj^r: Smith be discharged from being Overseer's of the Same, and Ordered y^e: Abraham: ffield be kept overseer of the said Roads (now Joyned) and that he follows the Directions of Maj^r: Augustine Smith (who is appointed Director) in Carrying on and Clearing the Same, & all ye tithables that belonged to both ye afores^d. Roads do help ye said ffield clear the Same --

1 March 1725/6 O.S., Page 103
On Petition of M^r: Robert Taliaferro for a road from Baylors Mountain to the falls, is granted, and Capt: Thomas Chew & M^r: Robert Taliaferro are appointed & ordered to veiw and lay ouf the same the most convenients way, and that Thomas Jackson be overseer of the said road, and it is further ordered that all the male tithables that are above Nicholas Christophers do help the said Thomas Jackson cleare the same --

5 April 1726 O.S., Page 109
Ordered that a Generall order doe Issue to all Surveyors of roads of this county that their roads be kept in good repair --

5 April 1726 O.S., Page 109
On motion of William Beverly Gent^n: he hath liberty to clear a road from his own plantation to the new road the most convenients way –

1 November 1726 O.S., Page 113
On Petition of Phillip Todd in behalf of himself & Severall others to have a road cleared on the south side of the litle mountains as far as the Hanover line is granted, and John Scott is appointed overseer viz: from ffox point to Hanover line, and all the tithables above Joseph Hawkins are ordered to help the said Scott cleare the Same, and all the

tithables below the said Joseph Hawkins are ordered to help Thomas Jackson (who is overseer below) to clear from ffox point to Taliaferros road –

2 November 1726 O.S., Page 116
On motion of William Beverly gent^n: to have the forke road as Abram Feild was overseer to be divided by the mountain Run is granted & ordered that William Russell Gent^n be overseer of the lower part and Robert Slaughter Gent^n: be overseer of the upper part, & the said Slaughter is ordered to continue the road to the ford that goes over Elk river in the litle forke above the said Beverlys plantation and that both gangs do meet and help make and keep the bridge in repair –

2 November 1726 O.S., Page 116
Robert Slaughter is excused of being overseer of the germana road, and Benjn: Cave is appointed overseer in his room--

2 November 1726 O.S., Page 116
Larkin Chew Gent^n: is appointed overseer of the road from the county bridge to Massauponax wharfe and the county road to Greens branch, and all the tithables below Edward Franklyns on the north side the river Po, including Stephen Sharps, /// Woodfordes quarter, /// Lawrance Smiths quarter William Tapps and all thereof of the inhabitants above Tapps are ordered to help the said Chew cleare ye same

2 November 1726 O.S., Page 116
Robert King is appointed overseer from the county bridge to the Church road and all the roads there abouts from the church road to the county line, & ordered that all the Inhabitants below the church and on the south side the River Po, to the river Ta are to help him cleare the same –

2 November 1726 O.S., Page 116
John Blanton is appointed overseer in the room of Charles Stephens decd: who was overseer of the roleing road from Nasauponax road to Mattapany main road, and to turn the road the most convenient way for a bridge into Massauponax new road, the former order of the old German road being made void --

2 November 1726 O.S., Page 116
Phillemon Cavenaugh with his gang is ordered to clear the road from the great road to Widdow Jael Johnsons fferry landing & to keep the same in good repair --

2 November 1726 O.S., Page 117
On petition of the Germans, Francis Kirkly and George James are desired and ordered to layout and marke the most convenients way for ye Germans mountain road, and Michell Holt is appointed Overseer of the same & all the Germans are ordered to help him clear the same --

2 November 1726 O.S., Page 117
John Waller is appointed overseer of the road from East North east bridge to the church on the river Ta, in the room of Dennitt Abney Junr and that he have liberty to clear with his gang the said road as it shall be thought convenient to have it altered for the better & laid of by Dennitt Abney & John Gambrell who are ordered & desired to to marke and layout the Same --

2 November 1726 O.S., Page 118 County Levy
To Larkin Chew $Gent^n$: for build y^e County bridge &
For Coroners Fee $\&^c$ as P his $Acco^t$... 9453 [lbs. of tobo.]

3 November 1726 O.S., Page 125
John Finleson being called according to Summons to, Answer the presentment of the Grand Jury, failing to Appear to Answer, for not clearing the road from the Wilderness bridge to Germana, Ordered that he be fined fiveteen Shillings to the use of his Majestie --

6 December 1726 O.S., Page 128
Henry Berry is appointed Overseer of the Road Viz^t: from Cowland to the bridge quarter that comes into Germana Road: in the room of William Wallers, & all the ththables between the said places are ordered to assist him clear the same --

6 December 1726 O.S., Page 128
Ambroes Grayson & e^c petitioning for a Road $\&e^c$ from Woodfords Quarter to Mr Francis Thorntons house is Rejected --

6 December 1726 O.S., Page 129
On Petition m^r: Francis Thornton to have the road that goes to the roleing house att the falls landing may be turned a small matter round by the mill, that one road may serve both mill and roleing house, is granted --

6 December 1726 O.S., Page 129
On complaint of Robert Taliaferro, against Thomas Jackson for not clearing the road and makeing a bridge as he ought where he is overseer, Ordered that he be fined fiveteen shillings according to Law, it being made appear P William Eddins

7 December 1726 O.S., Page 131
On petition of Thomas Phillips he is discharged from being overseer of the Road from ye Pine Stake to ambroes Madisons plantation to Capt: John Taliaferros Quarter and Nicholas Christopher is appointed in his room & ordered that all the tithables that lives about Capt: John Taliaferros quarter do help him cleare the same --

8 December 1726 O.S., Page 133
On Motion of John Taliaferro for to have a road from the pine Stake to his Quarter is granted & ordered that William Eddins & Richard Sharp, do lay of & marke the Same the most convenients way & least prejudicial to any person & make returne of their proceedings to the next court –

10 December 1726 O.S., Page 140
On petition of Mathew Noxum, (he being a Cripple) is excused from serveing on the high ways

10 December 1726 O.S., Page 141
On Petition of Richard Sharp for a road from the Mountain run to the end of Sharps path comeing into the roleing road, is granted and ordered that John Lee be Overseer of the same, and all the Inhabitants below William Eddins on the mountain run do help him clear the same –

3 May 1727 O.S., Page 146
On motion of Capt: Jeremiah Clowder, it is ordered that that road up Pamunky to the head of Mattapany branches (being not complyed with (where John Key is overseer) and being found (at present) not of much service) which is not cleared may be left of, and the road carried away towards the mountains as farr as pleasant run & that John Keys precints be from the place where the road is now cleared down to the next precints and that Jerimiah Clowder be appointed & made Overseer of the

upper precints, and all the tithables from Andrew Harrisons Quarter & John Bushs upwards to assitt him clear ye Same --

3 May 1727 O.S., Page 147
On motion of John Grame Gentn: about the Road from the mines to Masauponax It is ordered that Coll^o Alexander Spotswoods hands do clear from the mines to Germana road, and that all the tithables that served on that part of Germana road before, Law: Smith & Will Woodfords Quarter except [missing] do help clear the other part of the said road to Massauponax wharfe & that Thomas Germain be Overseer thereof --

3 May 1727 O.S., Page 155
William Eddins & Richard Sharp return their oppinion & vew of the road from the pine stake to m^r John Taliaferro plantation which was ordered to be lodged in the office and that the said Taliaferro have liberty to clear & make use of the same --

4 May 1727 O.S., Page 160
Ordered that a road be blazed laid out and a bridle way made at the County charge from Germana across the County to Northanna River, to goe near Capt: Jerrimiah Clowders or thereabouts for the use of the Inhabitants that live on that Side of the county, and that Capt: Jerimia Clowder be Impowered and desired to agree with some person to doe the same, which charge shall be paid at the laying the next county levy—

6 June 1727 O.S., Page 162
On Pettition of Edward Franklyn in behalfe of himselfe & Severall others for a road from the new german road where the road comes in that is called Franklyns road to m^{rs}: Jael Johnsons Ferry, It is ordered that Edward Franklyn George Procter and D^r: William Livingston or any two of them so lay of & marke the most convenienst way from the said place to the said ferry & make report of their proceedings to the next court --

4 July 1727 O.S., Page 174
On petition of Thomas Jackson he is excused from being Overseer of the lower part of the south side of the little mountains from Fox Point to Taliaferros road, and John Rucker is appointed to serve in his room, and all the tithables below Joseph Hawkings are ordered to help him clear the same --

4 July 1727 O.S., Page 175
On motion of Edwin Hickman gentⁿ: about rebuilding east North East bridge as the late great rains & Fresh had carried clean away, It is ordered that William Smith & edwin Hickman Gentⁿ: have power to agree with any workman to help and assist the said Edwin Hickman Gentⁿ: and his gang to build the said bridge, and that the county pay the charge of the said workman at the laying the next county levy --

4 July 1727 O.S., Page 176
On the Petition of Edward Franklyn for a road from the new German road, where the road comes in that is called Franklyns road to M^{rs}: Jael Johnson Ferry, the viewers appointed haveing made their report (who was appointed by the court to veiw and layout the same) Viz^t: Wee the subscribers have been and laid of and markt the road from the new German road to M^{rs}: Jael Johnsons Ferry, and wee find a plain streight ridge with but one small Hill which makes down to the low ground of the river June ye 23^d: 1727

                                              Edw: Franklyn
                                                      his
                                              George G P Proctr
                                                     marke

Which said road is granted & ordered that John Blanton be made & appointed Overseer thereof & that he the said Blanton with his gang do clear the same --

1 August 1727 O.S., Page 187
On petition of John Key, is discharged from being overseer of the highways from Rich neck upwards to the road that Jerimiah Clowder Gentⁿ: is overseer of and Anthony Goldson is ordered and appointed to serve in his room & the tithables that belong to John Key are ordered to serve under ye said Goldson, to aid and assist him clear the same (Andrew Harrisons quarter & John Bushes excluded) --

2 August 1727 O.S., Page 196
On Petition of Jerimiah Clowder gentⁿ: in behalf of himself and other Inhabitants on Pamunke To have a roleing road appointed and to be made & Cleared from ye ridge between Arseforemost & Plentifull to ye south side of Mattapony river, is Granted, and Ordered that Jerimiah Clowder Gentⁿ: be overseer thereof and that all the Tithables from Andrew Harrisons Quarter to John Bushes (inclusive) upwards Do assist him in Clearing ye same --

2 August 1727 O.S., Page 196
Ordered that Henry Berry Surveyor of Coll°. Corbins roleing road, take in the peice of a road from ye north side of Mattapony river & Joyn it to his precinct. And that all people that make use of ye bridge over ye river Ny (called corbins bridge) in roleing Tobacco Do when Occasions, help ye s^d. Berry and his gang to Keep ye said Bridge in repairs, and Likewise further Ordered that all ye male Tithables belong to Coll° Gawin Corbin do help & assist ye said Berry in makeing ye bridge over ye river Po --

2 August 1727 O.S., Page 196
On Petition of m^r. Robert Spotswood for ye removeing of Michell Holt from being Surveyor over ye Mountain road &c^a: and put in Michell Clore in his room, ye same is referred to ye next Court for Consideration—

7 November 1727 O.S., Page 207
On Petition of Henry Willis gent^n: for a road to be laid of from the main road att m^r: Beverlys ford in the fork of Rapahanock river up the little fork to his mill is Granted, and Ordered that Alexander Howard be overseer of the said road & that Maj Henry Willis tithables, and all the tithables below in the little fork and above Muddy run do help him clear the same –

7 November 1727 O.S., Page 210
The Grand Jury for the body of this County being Summoned & Sworn as the law directs & Received their charge from the Court Retired & the next day returned and brought in the following presentment Viz^t: Wee the grand Jury for our Sovereign Lord King and the body of this County do present John Finleson for not keeping the road in repair according to law from the land marke Stones below the bridge quarter to Germanna Abraham Feild foreman, It is therefore ordered that the said Finleson be fined fiveteen Shillings for the Same, except he appears at the next court & can shew cause if any, or give a reasonable excuse to the Contrary --

7 November 1727 O.S., Page 210
On Petition of Thomas Carr In behalfe of him Selfe and severall others for liberty to clear a bridle way from the fford that goes by the name of Devenports Ford to to the great road that goes up Arnolds run and a Cross the County to m^{rs}: Jael Johnsons Ferry, is Granted according to Petition --

8 November 1727 O.S., Page 210
On petition of William Thompson for a road from Maj^r: Henry Willis mill up to James Cannons is granted & ordered that William Thompson be overseer thereof and that all the Inhabitants above Maj^r: Henry Willis plantation do help & assist him clear the Same --

8 November 1727 O.S., Page 211 County Levy
To M^r: Benj^a: Cave for Marking out & Clearing a road as P order of court from Germanna to Northanna P Capt: Clowders and his order as P acco^t 1240 [lbs. of tobo.]

5 March 1727/8 O.S., Page 217
On Petition of Zachery Lewis in behalfe of him Self and severall others the Inhabitants of Mattapony & North Anna in this County to have a bridge built over the River Po att a ford called ffranklyns ford above Mattapony Church at the County Charge is granted, and Ordered that Larkin Chew, William Smith Edwin Hickman, and Henry Goodloe Gentlemen or any three of them who are appointed and desire to meet on munday the first day of aprill next being (the day before the next court) at m^r: John Snells plantation to agree with any one that will under take to build the said bridge at the said place (which person for ye Same according to Agreement to be paid P ye County att the next Laying the County Levy.) and that they make return of their proceedings to the next court and that the Sheriffe do give notice of ye same Accordingly--

5 March 1727/8 O.S., Page 217
On Petition of Richard Sharpe he is discharged from being overseer of the road from the wilderness bridge to the pine Stake towards the little mountains and Ordered that William Mackconico be Overseer in his room and agree to keep & clear the Said road, and the said hands that served under ye said Sharpe to help him clear the same --

5 March 1727/8 O.S., Page 217
On petition of John ffinleson he is discharged from being overseer of the road from the markd Stakes to Germanna & ordered that Thomas Dowdey do Serve in his room & with the said hands that did Serve under the S^d ffinleson do cleare the said road & keep the same in good repair --

4 June 1728 O.S., Page 233
On motion of Edwin Hickman Gent^n: he is discharged of being Overseer of the road from the County line to eastnor east bridge and M^r: Abraham

Abney is appointed overseer of the said road in his room and Ordered that all the male tithables that did Serve under m^r Edwin Hickman that live between the Said County line & East nor East bridge do help the said Abney clear the same --

4 June 1728 O.S., Page 233

M^r: Thomas Graves is appointed overseer of the road from East north East bridge to John Keys mill path & Ordered that all the male tithables that did Serve under Anthony Goldson that live between the Said bridge and Keys mill path do help him cleare the Same --

4 June 1728 O.S., Page 233

Anthony Goldson is appointed and continued overseer of the road from John Keys path that goes to Holidays mill, & So upwards to the road that Jerimiah Clowder is overseer of and ordered that all the male tithables that live between ye said Mill path & Capt: Jerimia Clowders road, (Andrew Harrisons & John Bushs plantation excluded) that did belong to him before, do help him clear the Same --

4 June 1728 O.S., Page 233

On motion of John Waller he is discharged from being overseer of the road from Mattapany Church to east north east bridge and the said road is devided into two precincts and John Wilkings and Daniell Brown are appointed Overseers in his room, Viz^t: from east north east bridge to John Wallers bridge includeing the same, m^r John Wilkings is appointed overseer & ordered that John Waller, Zachary Lewis John Wilkings, John Wiglesworth, Dennitt Abney sen^r: Dennitt Abney Junr: John Smith, Wm Dobbs, Daniell Pruett, m^r: Robert Baylors Quarter & Robert Stublefield, working Male tithables do help him clear & keep in repair the same --

4 June 1728 O.S., Page 233

and from the Said Wallers bridge to Mattapany church M^r: Daniel Brown is appointed overseer & Ordered that all the male tithables on the South Side of the River that belong to the Honble: John Robinson, Daniell Brown, John Micou, Abell Stears, John Naul, Patriack Bolding, W^m: Bradbourn, George Carters, Samuell Tillary, Phebe Hobson, & Robert Colemans plantions do help him clear the same --

2 July 1728 O.S., Page 236

On Petition of Larkin Chew Gent^n: he is excused from being overseer of ye road from the County bridge to Naussauponax wharfe, and from the County line to Greens branch, and the said road is devided into two precincts, & Ordered that William Tapp be overseer in the said Chews room on that part of the road from the County bridge to Nassauponax wharfe and that John Snell be overseer in the said Chews room on that part of the said road from the County line to Greens branch, and that all the male labouring tithables y[t] belong to m[r] John Lewis, The Hon[oble]: m[r] Grymes, Cap[t]: Larkin Chew, Cap[t]: Thomas Chew m[r] John Chew, and m[r] Richard Buckners plantations, do help the Said m[r] John Snell clear the same, and ordered that all the male labouring tithables on Rapahanock that did belong to the road before, do help the said William Tapp clear his part of the road as he is made overseer of --

3 July 1728 O.S., Page 237
On Petition of John Taliaferro Jun[r]: that the tithables on his plantation of this county, may be discharged from Serveing on the road that leads down Pamunky River as John Rucker was overseer, and be added & appointed to Serve on the road from the Pine Stake where Nicholas Christopher is overseer is granted & Ordered that they help the Said Christopher clear the said Road --

3 July 1728 O.S., Page 137
On Petition of John Bush for a road from the head of East north east to Mattapony Church in this county is granted, and ordered that Thomas Foster & Phillip Bush do view & layout the same the most convenienst way & least prejudicall to any ones land the said road goes through & make report of their proceedings to the next court & it is further Ordered that the said John Bush be made Overseer of the said road & that all the male tithables belonging to Richard Jones, Joseph Sheere, m[r]: James Garnett, m[r]: Nath: Sanders, Robert Andrews Phillip Bush, Robert Bush, Thomas Creders, & Joseph Roberts plantations do help him cleare the same --

3 July 1728 O.S., Page 237
Thomas Dowday is Discharged from being overseer of the road from the marked Stones to Germanna, and Benjamin Cave is appointed & ordered overseer in his room & that all the male labouring tithables that Served under m[r] John Finleson (the former surveyor) do now Serve under the said m[r] Benjamin Cave & help him clear the same --

3 July 1728 O.S., Page 237
Larkin Chew, William Smith & Henry Goodloe Gent[n]: made returne of the Agreement made with m[r] William Johnson & m[r]: Robert King upon building a bridge over at the River Po: as thay were appointed P the court: which Agreement is Viz[t]: In obeydiance to the within order wee the Subscribers have mett at the place within mentioned, and have agreed with m[r]: William Johnson & m[r]: Robert King to build a bridge about one

hundred yards above the said ford, to be built in a good Substantial & workemanlike manner ten feet wide, to cross Levil from the highest part of the point on the South Side the said river, to be done & finished by the last day of September next, to be built with good Substantial white oak plank. In consideration whereof, wee have agreed that they Shall have levyed at the laying the next County Levy two thousand pounds of Tobacco Cask & conveniency, Wittness our hands this 1st: day of Aprill 1728

                                        Larkin Chew
                                        William Smith
                                        H: Goodloe

likewise returned the Bond as the Said Johnson & King gave for two perform the said building: Which Said Bond & Agreement is Ordered to be kept in the Clerks office --

6 August 1728 O.S., Page 239
On Petition of John Snell to have more hands Added to his gang: is granted & ordered that all Cat^t: Harry Beverleys male labouring tithables: that did belong to Robert Kings gang be taken from him & added to John Snells gang to help him clear & keep his road as he is overseer of in good repair --

6 August 1728 O.S., Page 240
On petition of John Holladay in behalfe of him selfe & severall others for a road from his mill to the main road that goes P John Wallers above his quarter to the Church is granted & ordered that said John Hollady be made overseer thereof --

6 August 1728 O.S., Page 240
On motion of John Taliaferro Gent^n: it is ordered that m^r Stephen Sharps m^r Ambroes Graysons, William Perry, John Cammell, Harry Cammell, John Purvis George Purvis and Joseph Delany be added to his gang to help him clear the road as he is overseer -- & further ordered that Coll^o: John Grymes & m^r: Woodfords quarters that lies on Nassauponax do Serve under William Tapp to help him clear the road as he is made overseer ofe --

6 August 1728 O.S., Page 240
The Petition of William Eddins to have a bridge built over the Mine run, at the Countys charge, is Rejected

6 August 1728 O.S., Page 241
On complaint & Information made to this court P Benjamin Cave Overseer of the Road from the markt Stones to Germanna that John Grame Gent^n: Attorney to Coll^o: Alexander Spotswood will not suffer any of his male labouring tithables to come & help cleare the road with him pretending they are excused P a late act of assembly as belonging to ye Mines & c which if thay are taken away from him, he hath none other left to clear the said road that is now much out of repair, the same is referred to the next court for the consideration thereof, and to [blank in book] consider the law in that matter --

7 August 1728 O.S., Page 244
On Petition of m^r: Richard Cheek for a Road to Capt Larkin Chews mill P Lewis bridge &c is granted & Ordered that William Hansford Gent^n: & m^r John Snell do veiw the said road &c if it is necessary for one to be, if so then to layout the Same the most convenients way, & the least prejudicial to any ones land it goes through & make report of their proceedings to the next court—

8 August 1728 O.S., Page 253
On the petition of John Bush about a road from his house about the head of East north East to Mattapony Church & Thomas Foster & Phillip Bush being by the last court appointed to veiw and layout the same, & thay not haveing complied with the courts order, and Andrew Harrison preferring a petition for a road to the said Church alledgeing that the other road would be convenient only for the said Bush & m^r Garnetts Quarter, therefore on further consideration the Court have added, & Accordingly order that John Key & Daniell Brown with the said Thomas Foster & Phillip Bush or any three of them do meet & veiw th Said Lands & layout the most convenients way between them both to the said Church and the least prejudicial to any ones land the said road will goe through and make report of their proceedings to the next court --

8 August 1728 O.S., Page 253
On petition of William Johnson & robert King to have men appointed to veiw & Value So much timber as will be necessary to build a bridge over the river Po at the place known P the name of Franklyns Foard, as thay have undertaken to build on any man's land that lies is the most convenient, is granted & Ordered that Richard Blanton & Anthony Foster do veiw & value the same & make report of their proceedings to the next court --

3 September 1728 O.S., Page 254
On petition of Alexander Cleveland to have liberty to clear a bridle way from the church road into Germanna road is granted him & Ordered that he have liberty to clear the same --

3 September 1728 O.S., Page 255
On petition of Robert Green, John Roberts, Edmund Birk & Isaac Normon to have liberty to clear a roleing road from John Roberts P Robert Greens to normonds ford on the north side of Rapahanock river is granted them & Ordered that thay have liberty to clear the same --

4 September 1728 O.S., Page 258
On the motion of benjamin cave against John Grame Gent^n: attorney of Coll^o: Alexander Spotswood, for not Sending his tithables to help clear the road from the Markt Stones to Germanna, (being excused as the said Grame alleges P a late act of Assembly belonging to the Iron mines). Ordered that the said hands do Serve under & help clear the said road til such time as the Said Grame produces the law y^t he pretends does Excuse them--

4 September 1728 O.S., Page 258
On petition of Richard Cheek about a road to Capt Chews mill &c William Hansford & John Snell Gent^n: haveing made returne that thay do find the way that is petitiond for to be necessessary & accordingly have markt & laid it of as the order of Court directed: Therefore Ordered that Richard Cheek be overseer thereof And that all the male labouring tithables that belong to Elias Downs, William Hollaway, quarter, John Turner, Samuell Collins, Thomas Collins, Stephen Bickham, William Bickham and William Moore do help him clear the same --

4 September 1728 O.S., Page 258
On petition of William Johnson & Robert King about valuing timber to build the bridge over the River Po, Richard Blanton & Anthony Foster men appointed P the court to value the same haveing made returne of there proceedings of ye Valuation, Ordered that the Same be lodged in the Clerkes office --

5 September 1728 O.S., Page 264
On the petition of John Bush & Andrew Harrison for a road to be laid of from East north East to Mattapony Church, thay failing to appear when called in answer to their pettition Ordered that the same be dismist --

1 October 1728 O.S., Page 267
On motion of John Taliaferro Gent^n: about the Road from the South West Mountains to the Wilderness bridge, to be cleared P all the tithables from Thomas Downers at m^r John Taliaferro Junr: Quarter to George Wootons, and that William Moconico who being appointed Overseer formerly may be discharged and Benjamin Porter do Serve in his room as overseer of that part from the said Wilderness bridge to Nicholas Christophers and that the above Mentioned tithables do help him clear that part of the said Road accordingly -- is granted --

1 October 1728 O.S., Page 268 County Levy
to Mr William Johnson & Mr Robert King for building a bridge over the River Po as P Agreem^t wth ye Court --   2,000

To Ditto for Cask & Conveniency att 8 & 10 P Court of ye 2,000 --   360

2 October 1728 O.S., Page 270
On petition of Larking Chew Gent^n: in behalfe of Charles Chiswell & Comp^a: Gent^n: belonging to the Iron worke on Duglas run for a Road to be cleared from ye Intended place for the furnace to the most conveinients & nearest landing on Rapahanock River &c the furnace not being yett erected, the court are of oppinion at present that the said petition be Rejected –

2 October 1728 O.S., Page 272
On Motion of M^r: Robert Slatter he is discharged from being overseer of the Road from the Mountain run & the Said Road is devided into two precincts Viz^t: from the said Mountain Run to Bains Quarter Including Roger Oxfords & his tithables of which part m^r Francis Kirkley is appointed overseer thereof in his room and from Bains Quarter to M^r: William Beverleys fford, M^{r}r Robert Green is appointed overseer therof in the Said Slaughters room--

2 October 1728 O.S., Page 272
On motion of John Rucker he is Discharged from being overseer of the Road from Taliaferros road to Fox point bridge and Joseph Hawkings is appointed to Serve in his room as Overseer and all the tithables that belong to the Said Rucker do help the Said Hawkings & Ordered to Serve under him to cleare the Same --

5 November 1728 O.S., Page 274
On petition of Jerimiah Clowder Gent^n: he is Discharged from being overseer of the two roads Viz^t: from the head of the road by George Musicks to pleasant run the other from the ridge between arse formost and Plentifull to Mattapony --and M^r: Andrew Harrison is appointed to Serve in his room--

6 November 1728 O.S., Page 275
Our Petition of Charles Chiswell Gent^n: in behalfe of himself & Partners for a Road to be laid ofe marked and cleared from the intended place where they are prepareing to Erect a Furnace for makeing of Iron at a place called Fredricks Vill on a Run called Douglas Run in this county to Fredricksburgh Town nigh the falls of Rapahanock River is granted, And Ordered that m^r Edwin Hickman, Capt: Thomas Chew, m^r Andrew Harrison & m^r John Key or any three of them do Some time between this and the next court Marke and lay of the most Convenients & nearest way for a road as petitiond for, and make report of their proceedings to the next court --

7 November 1728 O.S., Page 283
Thomas Jermain is discharged from being overseer of the Nassauponax road and Henry Martin is ordered & appointed to Serve in his room--

7 November 1728 O.S., Page 283
On motion of m^r: Zachary Lewis to alter they way from Northanna to Fredericksburgh over the Hazell run, It is Ordered that William Livingston, Phillemon Cavenaugh & James Williams Gent^n: or any two of them do veiw the Same & make report of their oppinion about it the next court --

3 December 1728 O.S., Page 284
Richard Fitzwilliams Esqr: haveing made returne of the Viewers report that were appointed to veiw lay ofe & marke a road (on the pettition of Charles Chiswell & Caomp^a:) from the intended place where thay are prepareing to erect a furnace for makeing of Iron at Fredricksvill on Duglasses run in this county to Rapahanock river, none of the Gent^n: appointed that were at the laying of the Said road being present, to Inform about laying of the Said road into Severall precints for the more easey clearing the Same, the consideration of the Same is Refferred to the next court for the court to be Informed P Edwin Hickman Gent^n: who is one of the Court & one that was at ye laying of the Said road --

3 December 1728 O.S., Page 285
On Pettition of William Eddins for Satisfaction to be paid him for timber of his, as Benjamin Porter Overseer made use of to build the bridge of the mine run in this County is granted & ordered that Benjamin Porter, & Nicholas Christopher, do veiw & value the same as ye Said Overseer made use of & make report of their proceedings to the next court --

4 February 1728 O.S., Page 287
On the motion of m^r: Mosley Battaley for & in behalfe of John Hollady to have Some hands ordered to him to help him clear the road from his mill to the Church Road above John Wallers quarter that was granted P this court, it is ordered that all the male labouring tithables, of m^r George Seatons quarter, Thomas Sertains, John Sertain, Peter Gustavus and the S^d John Holladys own do help him clear the same –

5 February 1728 O.S., Page 288
On Petition of m^r Charles Chiswell & Comp^a: for a road from their intended Iron works on Duglas run called Fredricksville to Fredricksburgh on Rapahanock River, and the veiwers haveing returned that thay marked a road from the said Iron work to a landing on Rapahanock River at the mouth of the Hazel run between ye Lands of Coll^o: Man Pages & m^{rs}: Jael Johnsons to be the most conveniente Landing, which report was last court refferred to this for m^r Edwin Hickman one of the Said Viewers to Inform them, and m^r Zachary Lewis Attorney in behalfe of the Said m^r Cha: Chiswell & Comp^a: haveing put in a new pettition to have the said road & landing according as it is returned laid out & Marked, The Same is granted, and Ordered that m^r Thomas Jermain be Overseer of the Said Road & that all the male labouring tithables belonging to the Said m^r Charles Chiswell & Comp^a: and all adjacent Inhabitants that lives within two miles of each Side the Said Road So marked and laid ofe do help him clear the Same --

1 April 1729 O.S., Page 292
William Moore is appointed Overseer of the road Viz^t: from cowland to the bridge quarter that Comes into Germanna road, In the room of Henry Berry, and all the Tithables Between the said places are Ordered to assist him Clear the Same --

1 April 1729 O.S., Page 292
Richard Blanton is appointed Overseer in the room of John Blanton Who was Overseer of the roling road from Nassauponax road to Mattapony Main

road, and that All the tithables that Served under the S^d. John Blanton, Do help him ye S^d. Richard to clear the Same --

1 April 1729 O.S., Page 293
On petition of Charles Chiswell Gent. in behalf of himself & C°: his partners in the Ironwork at Fredricksville to have ye road (that was granted in February Court Last, from the Said Iron Work to a Landing on Rappahanock River at the mouth of the Hazle run Between the Lands of Coll°: Man Page and M^rs: Jael Johnson) Devided into three precints under three Distinct surveyors, Viz^t: The first precint from the place of the furnace to the ridge Between Parmunkey & Mattapony, the Second from the Said ridge to the River Ny, the Third from thence to the Mouth of the Hazle run on Rappahanock River And that the Labouring tithables within four Miles of the Said road in each of the precints afforesaid may be assigned for the making & Clearing thereof, And that m^r. Thomas Jarmain be Excused from being one of the Surveyors, the Same is granted & Ordered that Andrew Harrison be Surveyor of the first part Viz^t: from the place of the furnace of the Said Iron Work to the ridge between Pamunkey & Mattapony, (according as it is Laid out & marked P the Veiwers that were appointed) and that all the male Labouring tithables that are within four Miles of each Side of the Said road Do help him make and Clear the Same, And in Case of Sickness, removall or other Impediments of the Said Harrisson, It is further Ordered that John Key Do Serve as Surveyor in his room, And that two whom of the Said two, the Order Shall be Delivered to, the Same with ye people above Mentioned Shall make & Clear the Same–

1 April 1729 O.S. Page 293
That William Bartlett the Surveyor on the Second Precint of the Said road --Viz^t: from the ridge from Pamunkey & Mattapony to the river Ny, (according as it is Laid out & marked P the Veiwers appointed) and that all the Male Labouring Tithables that are Within four Miles of each Side the Said road Do help him make & Clear the Same --

1 April 1729 O.S., Page 293
That John Grayson Junr. be Surveyor on the third Precinct of the Said road Viz^t. from the river Ny to the Landing on Rappahanock river P the mouth of the Hazle run (according as it is Laid of and marked P the Veiwers appointed) And that all the male Labouring Tithables that are Within four Miles of each Side the Said road Do help him make and Clear the Same --

1 April 1729 O.S., Page 293
On Petition of Charles Chiswell Gent. in behalf of himself & c^a: his Partners in the Iron Work at Fredricksville to have bridges built at the County Charge over the rivers & Water Courses of the above Said Road the Same is rejected --

2 April 1729 O.S., Page 296
On Motion of Robert Slaughter Gent. in behalf of himself & others to have the former Way at the fflatt run quarter Cleared a gate to go through, It is the Oppinion of the Court that the road Do Continue as it is till the fall --

6 May 1729 O.S., Page 302
On Petition of edward ffranklyn to be Discharged from being Overseer of the road from ye Church on the River Ta to ffranklyns road on the head of Greens branch, is granted, An Samuel Hensley is appointed to be Overseer in his room, And Ordered that the Tithables which Served under ye s^d. ffranklyn Do help ye s^d. Hensley Clear the Same --

6 May 1729 O.S., Page 302 Grand Jury Presentments
Wee of the grand Jury being Sworn for the body of Spotsylvania County Do make Presentments #

We of ye Jury Pres^t. John Snell of the County of Spotsylvania and Pari^h. of S^t.George for Stoping the rode Between Rich^d. Cheeks and Lewis Bridge within this Six Months Last Past --

Wee Likewise Pres^t. John Snell & William Tapp both of the County of Spotsylv^a. and Parish S^t. George for not keeping the bridge over Warners River according to Law within the Six Months Last Past --

6 May 1729 O.S., Page 307
On the petition of William Eddins for Satisfaction to be paid him for Timber made use of in Building the bridge over the Mine run, the Same being referred to Value and appraise the S^d. Timber, And the Valuation being returned amounting to four hundred pounds of Tobacco P ye viewers appointed, Which ye Court thinking to be too Extravagently appraised, Are of oppinion and accordingly Order that he be paid One hundred & fifty pounds of Tobacco and no more for Y^e s^d. Timber by the County at the laying the Levy --

7 May 1729 O.S., Page 308
On Motion of M^r. Zachary Lewis about having veiw'd y^e. best Way to go over the Hazell run the Order of Court not being Complyed With or returned, Ordered that Philemon Cavenaugh Overseer of the Said road With his gang Do make a bridge over the said Hazell run as the road now goes --

3 June 1729 O.S., Page 316
On Petition of John Bush to have a bridge made over the river Po, in the roling road Where Capt: Jerimiah Clowder is overseer of, the Same is rejected --

3 June 1729 O.S., Page 316
On petition of Michael Clore to have the road Cleared from M^r: John Lightfoots Plantation into Germanna road, and to have more hands added to their Gang with another Overseer, is granted, And is Ordered that Christopher Zimmerman be Overseer thereof, and that Joseph Bloodworth, Joseph Fox, Frederick Cobler, David Jones, Joseph Cooper, and Conred Ambergue their tithables be added, to aid and assist the Said Clore and his gang to Clear the Same --

3 June 1729 O.S., Page 316
On Motion of M^r: William Russell, It is Ordered that M^r: Robert Green & gang, M^r. ffrancis Kirkleys Gang, With his own Gang, Do help Make and build the bridge Over the Mountain run, and keep the Same in good repair --

3 June 1729 O.S., Page 319
John Snell appearing to answer the presentment of the grand Jury according to Summonds for Stoping the road between Richard Cheeks and Lewis bridge Within this Six Months Last past the Court having heard all arguements and the Said John Snells oath &c, are of oppinion, that the said presentment is not good & ordered y^t. the Same be Dismist --

4 June 1729 O.S., Page 321
On Petition of John Bush for himself and the rest of the Inhabitants of the Glady fork for a road from Capt. Jerimiah Clowders road to Mattapony Church over the bridge already built, is granted, and Edward ffranklyn is appointed overseer thereof, And Ordered that all the male Labouring tithables belonging to John Durrett, Walter Butler, Nathaniel sanders Quarter, James Garnets Quarter, Joseph [blank in book], Richard Jones,

Edward ffranklyn, Phillip Bush and John Bush, Do help the said ffranklyn to Make and Clear the Same --

4 June 1729 O.S., Page 324
John Shell and William Tapp appearing When Called according to Summonds to answer the presentment of the grand Jury for Not keeping the bridge over Warners River according to Law within Six Months Last past, and the Court having heard their arguements and Excuses Are of oppinion that they be Excused and accordingly Order that ye s^d. presentment be Dismist

1 July 1729 O.S., Page 325
On Petition of Edward ffranklyn about altering the road (which John Bush obtained Last Court from Capt. Jerrimiah Clowders road to Mattapony Church over the bridge already built) in having the Same goe over about Seventy Or Eighty Yards above ye s^d. bridge, Ordered that John Durrett & Walter Butler Do veiw & Layout the Same and Make return of the best Way to the Next Court --

5 August 1729 O.S., Page 330
On Petition of John Snell to be discharged from being overseer of the road from the county line to the head of Greens branch (haveing served thirteen months) is granted, and ordered that Nicholas Hawkins, do serve as overseer in his room--

5 August 1729 O.S., Page 331
On Petition of Michael Holt in behalfe of him self & others for liberty to clear a road into the main road from the Island in the first fork of the white oak run for to role their Tobacco is granted & Ordered that thay have liberty to clear the same --

5 August 1729 O.S., Page 331
On Petition of Charles Stephens in behalfe of him selfe and others for liberty to clear a Road from Cat^t: Jerimiah Clowders roleing road to m^r Augustine Moores quarter in ye ffork of Pamunky is granted & Ordered that thay have liberty to clear the same –

6 August 1729 O.S., Page 335
On Petition of M^r: Charles Chiswell & partners about having bridges built over the River Po & Ny: at the Charge of the County, the Court having Considered the same, Ordered that y^e. Severall Overseers Do forthwith Put & keep the old road in good repair, And that Thomas Chew, Henry Goodloe, Joseph Brock & William Johnson Gent^n. or any three of

them Do Likewise forthwith Meet and agree with Some Workman to build a bridge over the River Ny at the County Charge, And they are Desired to Veiw the Nighest & most Convenient Place to make a bridge over the River Po: for the use of the said Mine Company & Make report of their Proceedings to the Next Court --

6 August 1729 O.S., Page 335
On Motion of Benjamin Cave for More assistnce to help him keep his road in good repair (John Grame Gent^n. refuseing to Send Coll^o: Alexander Spotswoods Tithables on the said road Pretending they all now belong to the Mines) It is Therefore Ordered that The severall Tithables belonging to the Sundry People hereafter Mentioned Viz : Majr: Goodrich Lightfoots, George Wheatleys, Thomas Stantons, John Bond, George Smiths, John Smiths, Thomas Parkes, Philemon Cavenaughs Quarter, John Gordons, Coll^o. Gawin Corbins four Tithables at his Quarter, Mr. John Grame's and Coll^o. Alexander Spotswoods four tithables at his Chesnut Quarter (Provided that if the s^d. John Grame gent. Send the same Number of Tithables belonging to Coll^o. Alexander Spotswood then ye s^d. four Tithables belonging to ye s^d. Chestnut Quarter to be Exempted from Comeing on ye s^d. road) be added & Exempted from all other roads & that they help the said Cave Clear y^e. said road and Keep the same in good repair –

6 August 1729 O.S., Page 336
Ordered that M^r. William Bledsoe & M^r. Benjamin Cave Do value the Timbers the Overseer Do Make use of in building the Wilderness Run bridge, And that they make return of their proceedings to the Next Court --

6 August 1729 O.S., Page 336
Ordered that M^r. William Russell & all his Tithables that Serve under him Do mend the bridge at the fountain & Clear y^e. Said road up to the Court house & And keep the Same in good repair --

6 August 1729 O.S., Page 336
On Petition of Edwin Hickman & Zachary Lewis in behalf of themselves & others to have Liberty to Clear the Nighest & most Convenient ^{W}ay from Arnolds run to Germanna, Ordered that they have Liberty to Clear a bridle Path, takeing a way no Persons Imediate Conveinences --

6 August 1729 O.S., Page 336
On Petition of Joseph Hawkins about having the South West Mountain Road Devided into Two Precints, is granted, & Ordered that the same be

Divided from Craffords Tomb Stone, And that Abram# Bledsoe be Surveyor of that part, And that all the Tithables Below Capt: Thomas Beals Quarter Do help the said Bledsoe Make Clear & keep in good repair the Same, And It is further Ordered that all the Tithables belonging to both y^e. precints do help make y^e. bridge & keep the Same in good repair --

2 September 1729 O.S., Page 340
On petition of John Wilkings, he is discharged from being overseer of the road from Wallers bridge to East north east bridge, & Ordered that John Smith do Serve in his room & that the same tithables which Served under ye Said Wilkings, do help the said Smith cleare & keep in good repair the said road

2 September 1729 O.S., Page 340
On motion of m^r William Russell that m^r Henry ffeilds tithables may be added to his gang to help clear ye road, the same is Rejected --

2 September 1729 O.S., Page 340
On petition of William Tapp, he is discharged from being overseer of the road from the county bridge to Massauponax wharfe, and ordered that Bartholomew Wood do serve in his room, and that all the tithables that Served under the said Tapp do help the said Wood clear & keep in good repair the said road --

2 September 1729 O.S., Page 340
On petition of Henry ffeild to have his tithables discharged from Serveing under George Wheatleys road, & to be added to m^r William Russell gang to help clear his road, (being the most convenient,) the same is granted --

2 September 1729 O.S., Page 340
On Petition of John Taliaferro Junr: that the road near Nicholas Christophers may Extend up to the rockey run for the conveniency of the inhabitants above same is granted & Ordered that m^r Benjamin Porter & his gang do clear the same --

2 September 1729 O.S., Page 341
On Petition of Henry Willis & Phillimon Cavenaugh in behalfe of themselfs & others to have a road from the Said Willis' mill to the Courthouse over muddey run, is rejected.

2 September 1729 O.S., Page 341
On petition of William Davis James MackCullough &^{c} to have an overseer appointed in the room of John Lee who is gone out of the county of the road commonly called the county mountain road &^{c}, for the use of their plantations to goe into the Said mountain road, is granted & Ordered that James MackCoullough the overseer thereof & that the said Petitioners William Davis James M^c:Cullough Dennis Lindsey & george Waller with William M^c:Conikos & Richard Sharps tithables (where ye s^d Sharp now lives) do help him clear the same --

2 September 1729 O.S., Page 343
On Motion of John Scott Gent. to have the Mountain road Devided from Capt. Thomas Chews mill to the County Line, is granted, And Ordered that John Mynor be Overseer thereof & that all the tithables from above Thomas Beals Quatter Do help said Mynor to Clear & keep the Same in good repair --

3 September 1729 O.S., Page 344
On the Petition of Edward ffranklyn about altering the road from Capt. Jerimiah Clowders road to Mattapony Church, the said Petition is Ordered to be Dismist --

3 September 1729 O.S., Page 344
Ordered that Edward ffranklyn be Overseer of the road from Capt. Jerimiah Clowders road to Mattapony Church along the ridge by Cranwells y^e: Nearest and best way to the said Church, And that all the Tithables that served under him before Do help him Clear the same --

3 September 1729 O.S., Page 345
On Motion of edward ffranklyn M^r. robert Baylors Tithables at his Quarter, Thomas Creders & Joseph Roberts Are Ordered to be added to his gang --

3 September 1729 O.S., Page 348
On the Petition of Charles Chiswell Gent & Partners & c^a. for bridges to be built over the Rivers &^{a}. at ye County Charge, the Gentlemen appointed by the Last Courts order having not fully Comply'd with y^e. same, Ordered that the said Order be Continued for the s^d.Gentlemen to Compleat the same, and make report of their Proceedings to the Next Court --

3 September 1729 O.S., Page 348
On Petition of Benjamin Cave he is Discharged from being Overseer of the road from Germanna to the Marked Stones, and Ordered that John Gordon do Serve in his room, and that the Same Tithables which Served under the said Cave Do help the said Gordon Clear and keep in good repair the said road.

3 September 1729 O.S., Page 348
On Petition of Benjamin Cave for & in behalf of himself and others for a road from the Walnut Branch on y^e. North Side of Rappadan Down the Ridge & to Cross y^e. Rappadan so Down to the South West Mountain Chappell, is granted, And Ordered that the said Benjamin Cave be Overseer thereof, And that all the Tithables which are Situated adjacent on y^e. River Do help him ye s^d. Cave Clear the same --

7 October 1729 O.S., Page 352
On Petition of anthony Goldson for more hands to be added to his gang on the road that he is Overseer of, the same is rejected --

7 October 1729 O.S., Page 352
Phillemon Cavenaugh Overseer of the Rappahannock Road Petitioning this Court for order and Directions Where he may have Timber to build the bridge Ordered to built over the Hazle run (P reason M^r. James Williams Owner of y^e. Land adjacent have forewarned him to fall any trees of his) It is Considered P the Court that their is no great Occasion for building that bridge, Ordered that the said Petition be Dismist --

7 October 1729 O.S., Page 353
On Motion of Benjamin Porter Overseer of the South West Mountain road for Order and Directions Where he may have Timber to build the bridge over the Mine run (wm. Eddins having forewarned him from falling any trees of his) It is Ordered that ye said Benjamin Porter have Liberty to agree with any person Living the most Convenients to ye s^d. Bridge to buy timber to build and repair the same and Cheapest and Best Terms he Can, and they to be Paid for ye s^d. Timber by the County at the Laying the Next County Levy --

7 October 1729 O.S., Page 353
On ye Petition of Mr. Charles Chiswell & Partners about having Bridges over the Rivers Po & Ny at the County Charge, the Gentlemen Appointed having made return & ye Agreement made about building that bridge over the River Ny, but the other part of the said order not being Complyed

With, Ordered that the Same be Continued to the Next Court for the Gentlemen appointed to fully Comply with the same & make return therof --

7 October 1729 O.S., Page 354 County Levy
To John Wiglesworth for building the Bridge over the River Ny, according to agreem^t 2,000 --
....
To John Waller ass: of W^m. Eddins for Timber to mend the Mine bridge & Copy Order   158 --

7 October 1729 O.S., Page 355
On Petition of John Gordon Overseer of the road from the Markt Stones to Germanna, for to have the Wilderness run bridge built at y^e. County Charge (P reason his gang Chiefly Consisting of Coll^o. Alexander Spotswoods Mine People Which now are Exempted P Law) is granted, and Ordered that Goodrich Lightfoot William Bledsoe, and Robert Green, Gentlemen or any two of them, do agree with Some Person to build the same on the Cheapest Terms they Can and Make return of their Proceedings to the Next Court --

7 October 1729 O.S., Page 355
Ordered that all those Tithables Which Live in ye fork be Discharged from Serving under John Gordon Overseer of the road from the Markt Stones to Gerrnanna, P reason the bridge now being to be built on the County Charge --

4 November 1729 O.S., Page 356
On petition of Edward ffranklyn to be discharged from being overseer from Capt: Jerrimiah Clowders road to the church on the river Ny is granted and Ordered that John Durrett do Serve in his room--

4 November 1729 O.S., Page 356
On petition of Abraham Abney he is discharged from being overseer of the road from the county line to East North East bridge and Ordered y^t George Woodrofe do Serve in his room--

5 November 1729 O.S., Page 358
On motion of m^r: Zachary Lewis in behalfe of m^r Charles Chiswell Gent^n: and diverse others adventurers in an Iron works upon Duglas run in this county to have the order of the Generall court bearing date October y^e

24th: 1729: recorded --and that men may be appointed to veiw and value all such timber as Shall be necessary for the building a bridge over the River Po According to y^e. s^d Order, is granted & Ordered that John Holladay, John Key, & Robert King or any two of them do appraise all such timber when the same shall be cutt down fore ye building the said bridge and make report of their proceedings to y next court

The Copy of ye Order of the Generall Court Viz^t: --

At a General Court held at the Capitol, October the 24th: --

Charles Chiswell Gent in behalf of Himself and Divers others adventurers in an iron work upon Duglass run in Spotsylvania County Called ffredericksville this Day produced an order of the Court of the said County dated the fifth of ffebruary last for Laying out and making a new road from the said Iron Work to a Landing on Rappahanock River at the mouth of the Hasle Run Between the Lands of Man Page Esqr. and M^{rs}.. Jael Johnson according to the report of Veiwers appointed by the said Court for that purpose persuant to an act of assembly lately made and an order of the said Court made the first Day of April ffolowing for Dividing the s^d.. road into precincts and appointing Surveyors for the same and also another order of the said Court Dated the Sixth of August Last for Preparing the old road and building bridges over the Rivers Ny and Po at the County Charge and an order for Continuing the same made the seventh of October following whereby as he Suggested the former orders were by the Inhabitants of the said County Supposed to be sett aside and therefore moved that the said new road being in a Great Measure already Cleared and a bridge being made over the river Ny might be Cleared and Compleated as being most Convenient for the said Iron work and that a Convenient Bridge might be built over the river Po at the said Counties Charge, And the Court having heard the said Chiswell and also Joseph Brock on behalf of the said Court thereupon Do order that the order of the said County Court made in August and October Last be set aside and that the said orders made in ffebruary and April last be Confirmed and also that the road therein mentioned be Continued and Established as the road from the said Iron work to the said Landing on Rappahannock River and be Cleared and Compleated as soon as the same Can be Conveniently Done and that the said Chiswell have Leave at his own Expwense to build a Substantial and Convenient ridge over the said River Po and the said County Court at the next Levy are to allow him his reasonable Charges and Disbursements about the same and in the meantime to appoint fit persons to appraise the timber which shall be Necessary for the building the said bridge and the same Shall be Cutt Down to the End the owners thereof may be duly Satisfied, And for as much s it is of light disputed whether Surveyors of the Highways are by Law Impowered to take wood in the adjacent Lands for the making and repairing Such bridges and Causewys as they are obliged to make and repair whereby the Due execution of the Laws Concerning highways is Likely more and more to be obstructed this Court have thought fit to Declare their opinion that Surveyors of the Highways may Lawfully take such wood for their purpose as is nearest and next adjacent to the bridges and causewys which are to be made and repaired by them so as they ct therein with Caution and a Strict regard to the Interests of the owners of such wood and do the least Damage that Can be to them--

Cop: Test Jn°: Frauncis DCGC

5 November 1729 O.S., Page 361
The last courts order which appointed Goodrich Lightfoot, William Beldsoe and Robert Green Gent^n: to agree with Some workman to build a bridge over ye Wilderness run &^{c} at the county charge, not being complyed with, the same is continued to the next court to compleat the same & make report --

2 December 1729 O.S., Page 362
On Petition Daniel Browne in behalf of himself & severall others appointed to Clear the Midle Precints of the Mine road from ffredricksville to Rappahanock river &c^a: Setting forth that they not being above twenty five tithables in the Said Precints are not Strength enough to Clear the Same and Make Bridges and Causways as is required to be Done, Praying that more help may be assigned them to help Clear the Same &c^a: is rejected --

2 December 1729 O.S., Page 362
On Petition of Henry Martin to be Discharged from being Overseer of the Nassauponax road, the same is granted, And Ordered that Charles Steward Do Serve as Overseer in his room --o

2 December 1729 O.S., Page 362
On Motion of Harry Beverley Gent. against William Moore Overseer of the road from Cowland to the Bridge Quarter that Comes into Germanna road for not Keeping ye s^d. bridge and Grubing the Said road &c^a: as the Laws Directs Which being made appeared to the Court, Ordered that he be fined fifteen Shillings for the Same with Costs, therefore Ordered that he Pay unto y^e. said Harry Beverley Gent. the Informer the Same alias Execution --

2 December 1729 O.S., Page 362
On Petition of William Russel to be Discharged from being Overseer of the road from Germanna to the Mountain run bridge in the fork of

Rappahanock river, the Same is granted, and Ordered that Mr Samuel Ball do serve as Overseer in the room of y^e. s^d. Russel --

2 December 1729 O.S., Page 362
On Motion of Coll: Henry Willis to have a road from his Mill in y^e. of Rappahanock River to Germanna, the Same is granted, And Ordered that Robert Green, William Russell, and ffrancis Kirkley Gent. Do Veiw, Lay out, and Mark, the Nearest & best Way, And Make return of their Proceedings to the next Court --

3 February 1729 O.S., Page 370
On Petition of William Moor he is Discharged from being Overseer of the road from Cowland to y^e. bridge quarter that Comes into Germanna road, and Ordered that Joseph Williams Do Serve as Overseer in his room, and that all the Tithables Which served under ye s^d. Moor, Do help the said Williams to Clear the Same --

3 February 1729 O.S., Page 370
On Petition of Charles Stevens and Severall others to have the road that Liberty Was given them to Clear from M^r: Augustine Moore's Quarter in the fork of Pamunkey to Capt: Jerimiah Clowders roling road, Drop, that not Suiting the Inhabitants and that Capt: Jerimiah Clowders roling road may be Continued up the ridge in the S^d. fork. of Pamunkey the Most the best and Convenients Way towards y^e. Mountains as farr as the People in those Parts are inhabitted is granted, and Ordered that John Cook, Charles Stevens, David Cave, John Henderson and Thomas Cook have Liberty to Clear the Same --

4 February 1729 O.S., Page 372
The Veiwers appointed to Veiw & Mark out the nearest and best way from Coll^o: Henry Willis mill in the fork of Rapahannock river to Germanna, haveing made return of the said order which the court considered & ordered the Same be Dismist --

4 February 1729 O.S., Page 373
On motion of m^r John Mercer in behalf of William Bartlett Overseer of the midle precincts of the mine road (that is granted & laid ofe from ffredricksvill to Rapahanock river below the Hazell run) and the gang that is under him, Setting forth that each other of the precints are shorter and treble the number of hands to clear it, especially the first part that is between ye Mine & ye midle precint does not exceed Six miles in length, & their precints is about ten & all to new clear and Many Small bridges & long casways to make, It is therefore ordered that

John Key with his gang do clear two miles in length more towards & in the midle precincts & that John Grayson with his gang do clear one mile further from the bridge built over the River Ny into the Midle Precint of the Said road --

5 February 1729/30 O.S., Page 376
On motion of Goodrich Lightfoot Gent^n: in behalfe of him selfe & Severall others to have a bridle way from the ferry att Germanna into the road that comes P Coll^o: Alexander Spotswoods old mill, is granted & Ordered that George Wheatly & the gang of hands under him do clear the same --

5 February 1729/30 O.S., Page 377
On petition of Anthony Goldson he is discharged from being overseer of the highway from John Keys mill path to cap^t: Jerimiah Clowders part of that road and Thomas Pulliam is ordered to Serve in his room & all the hands that were under the Said Goldson is ordered to Serve under & help the Said Pulliam clear the same --

5 February 1729/30 O.S., Page 377
On motion of Harry Beverley Gent^n: against Nicholas Hawkings overseer of the road from the county line to the head of green branch for not clearing & keepjng his road in good repair according to law, it being read & appear to the court, Ordered that he be fined as the Law directs for the Same --

3 March 1729/30 O.S., Page 379
It is ordered that the main road P Rapahanock river be altered & cleared & the same be made and cleared on the back line of Fredricksburgh town the most conveniens way for the good of the said town & others, & that Phillemon Cavenaugh Overseer with his gang that is under him do alter and cleare the same according to the direction of Henry Willis Gnet^n: --

3 March 1729/30 O.S., Page 381
On petition of Mr Zachary Taylor for liberty to turn and alter the road (that M^r: Benjamin Cave obtained an order of this court to clear through the Southwest Mountains as he have marked in that part of his land where he is goeing to build) about two hundred & fifty yards to come into the Mountain road a little higher, the Same is granted & ordered that he have liberty to turn and alter the Same, according to petition --

3 March 1729/300 O.S., Page 382
On petition of Nicholas Hawkings to be discharged from being overseer of the road from the County line to the head of Greens branch he being removed out of that precinct, is granted and Ordered that m^r Rice Curtis do Serve in his room & all the tithables that did Serve under the S^d Hawkings, do now Serve under & help the Said Curtis to Clear & keep the said road in good repair --

4 March 1729/30 O.S., Page 384
On the petition Exhibitted P Mr Zachary Lewis in behalfe of Mr Charles Chiswell & the Iron mine Comp^a: to have the road of the midle precincts confirmed where the Overseer varied from the line as the veiwers made report of, the Same is Rejected --

7 April 1730 O.S., Page 385
On Petition of John Currett to have the road as he is Overseer of from John Bush's to Matapony Church Devided into two Precints the Same is rejected --

7 April 1730 O.S., Page 385
On Petition of Thomas Pulliam Who Was appointed Overseer in the room of Anthony Goldson of the road from Keys Mill Path to Capt. Jerimiah Clowders Park of that road, for Tithables to help him Clear & repair the same is granted, And Ordered that Anthony Goldson, Paterson Pulliam, Robert Turner and Coll^o. Moore's Tithables at his bridge Quarter, Do help him Clear & keep in good repair the Same --

7 April 1730 O.S., Page 385
On Petition of William Bartlett Overseer of the Midle Precints of the Mine road, (Viz^t.from the ridge Between Pamunkey and Mattapony to the river Ny According as it is Laid out & Marked P the Veiwers appointed) to be Discharged from the Same, is granted, And Ordered that Richard Blanton Do Serve as Overseer in his room, And that all the Male Labouring Tithables that are Within four Miles of Each Side the said road help him make, Clear & keep in good repair the Same --

7 April 1730 O.S., Page 386
On petition of Coll^o: Henry Willis to have a road to his mill in the fork of Rapahannock River from m^r John ffinlesons road to the upper inhabitants & from the Said road to his mill & is granted & Ordered that Jonas Jenkings John Ashby & William Smith, or any two of them do view &

layout the most convenients road According to the Said petition & make report of their proceedings to the next court --

7 April 1730 O.S., Page 387
M^r: John ffoster Deputy Sheriff returned the Courts order Wherein they appointed Rice Curtis to be Oversser of the road from the County Line to the head of Greens branch, and Informed the Court that the said Curtis Would not Accept of the Said Order to Put it in Execution (Which he Declared on Oath) Ordered that the Said Curtis be fined twenty Shillings Current money for his Contempt and that he be Continued Overseer of the Said road --

5 May 1730 O.S., Page 388
Richard Blanton is appointed Overseer of the Precints of the Mine road in the room of William Bartlett, And Ordered that ye s^d. Blanton With the Tithables Which Served upon the Said road under the former Overseer, keep in good repair the Same as the Said Bartlett have already Cleared --

5 May 1730 O.S., Page 389
On Petition of Thomas Crethers, Joseph Roberts George Musick, to be released from Serving on five roads, is granted, And Ordered that they Only Serve on the Mine road and the road that goes from Capt. Jerimiah Clowders roling road to the Church--

5 May 1730 O.S., Page 389
On Motion of M^r: Zachary Lewis in behalf of M^r. Rice Curtis to be Discharged from being Overseer of the road from the County Line to the head of green's branch, as Nicholas Hawkins Was formerly Overseer of, is granted, and Ordered that Robert Hutcheson Do Serve in his room (he being freed and Exempted from all other roads) and all the Tithables that Served under the former Overseer Do help the S^d Hutcheson Clear and keep in good repair the Same --

5 May 1730 O.S., Page 389
M^r: Rice Curtis Appeared and Moved that the Court Would remitt the fine of twenty Shillings As he Was Mulcted Last Court for not Accepting of the Order of Court, and Clearing the road as he Was Appointed Overseer of, and that they Would be Pleased, to put Some other Person in his room, he giving the Court Satisfactory reasons, It is Ordered that he be Discharged from being Overseer of the Said road and that Coll^o. Henry Willis be Desired to Wait on his Hon^r.The Governour to Interceed to have the Said fine remitted --

5 May 1730 O.S., Page 389
John Vinton is made Overseer of the road Viz^t: from the falls of the Wilderness bridge And Ordered that all the Tithables that Serve under the former Overseers of the Said road Do help the Said Vinton Clear and keep in good repair the Same --

7 July 1730 O.S., Page 402
John Wiglesworth and Richard Blanton Came into Court and gave bond for Keeping the Mine bridge that The Said Wiglesworth built and finished Over the River Po, in good repair Seven Years According to agreem^t: Made with Mr Charles Chiswell, Which was Ordered to be Lodged in the Clerks office --

7 July 1730 O.S., Page 402
On Petition of Benjamin Porter to be Discharged from being Overseer of the Mountain Road, the Same was rejected --till such time he hath made the said road in good repair --

7 July 1730 O.S., Page 403
On the Petition of Henry Willis Gent^n: about a road to his mill, the Same is continued to the next court, he being on the Assembly--

4 August 1730 O.S., Page 404
On Petition of Robert Hutcherson Overseer of the road from the County Line to the head of Green's branch, Which Said road Cross's y^e. bridge over y^e. River Ny, and that S^d. bridge being so out of repair that his gang Was not able to Mend & repair the Same Neither is there any Timber to Do it, the Same is rejected till the New Law be Sent to this Court & Ordered that y^e. Clerk Do Issue out an Order in the Intrim to him for all the Tithables y^t. Served under Nicholas Hawkins former Overseer to help y^e. s^d. Hutcheson repair the said bridge --

5 August 1720 O.S., Page 414
On the Petition of Henry Willis Gent. about a road to his Mill in the fork of Rappahanock river, the Last Court's order appointing Veiwers, and the Veiwers having not made their return about the Same, Ordered that the Petition be Continued --

1 September 1730 O.S., Page 420
On the Petition of Col^o. Henry Willis to have a road to his Mill in the fork of Rappahanock River from m^r. John ffinlesons road to the upper

Inhabitants and from the said road to his mill & c^a., the Veiwers appointed having made their report, On the Said Willis's Motion and Order is granted for a reveiw, and Ordered that M^r. ffrancis Slaughter ffrancis Kirkley & William Payton Do Veiw the Same, and that the Surveyor of that Part of the County with the aforesaid three Gentlemen Do Lay of y^e. Same the Most Nighest & Conveyants Way (all the Said Willis's Charge) and make report of their Proceedings to the Next Court --

2 September 1730 O.S., Page 421
Ordered that John Holladay, John Key, Robert King and John Wiglesworth or any three of them whereof the said John Wiglesworth Shall be one, (being first Sworn before a Majestrate of this County) Do Value the Timber made use of in building a bridge over the River Po on y^e. Mine road, and Unto whom ye said Timber Did belong to and Make return of their Proceedings to the Next Court --

7 October 1730 O.S., Page 429 County Levy
To Mr. Charles Chiswell for Paid John Wiglesworth for Makeing the bridge over the River Po…..   3000

7 October 1730 O.S., Page 430
Ordered that Daniel Brown, Richard Blanton, William Bartlett & John Wiglesworth of any two of them (being first Sworn before a Majestrat of this County) Do Value the Timber Made use of in building a bridge over the River Po in the Mine Road In Tobacco And unto Whom y^e. s^d. Timber Did belong to, and Make return of their Proceedings to the Next Court --

Spotsylvania County Order Book 1730-1738

3 November 1730 O.S., Page 1.
On Petition of John Gordon Overseer of the road from the Markt Stones to Germanna to be Discharged from the Same, (he being removing from that Precint) is granted, And Ordered that Thomas Farmer Do Serve in his room --

3 November 1730 O.S., Page 3   Grand Jury Presentments
    Wee Likewise Par^t. John Gorden Overseer of the rode of this Parr: & Co^{ty} afoursaid for not keeping his Part of Rode in repair Within this two months Last Past
* * *
Wee Likewise Par^t. Robert Hutcheson of the Co^{ty}: and Parr. aforesaid for not Keeping Lewis Bridge in good repair Within this two months Last Past --

4 November 1730 O.S., Page 4
On Petition of Andrew Harrison he is Discharged from being Overseer of the road from the ridge between Arseforemost and Plentifull to Mattapony, And Thomas Credders is appointed, And Ordered to Serve in his room, And all the male Tithables below M^r. Beverleys Quarter are Ordered to Serve under the Said Credders & help Clear the Said Road, And that the road from George Musicks to Pleasant run be Continued the Nearest and best way to the Mountain road, and that the Said Harrison be Continued Overseer of the said road, And all the People above the said Beverleys Quarter and including ye s^d. Beverleys Quarter as formerly Did belong to his gang Do help him Clear the Same --

1 December 1730 O.S., Page 8
On Motion of Maj^r: Augustine Smith to have the old road by the fflat run Quarter Opened and Cleared as Were formerly Cleared & made use of, (Which Was objected against by John Grame Gent. Who Desired a Veiw, Which was Overruled) is granted, and Ordered That Thomas Farmer Who is appointed Overseer of the Said road in the room of John Gordon, With his gang Do Open and Clear the Same --

1 December 1730 O.S., Page 8
On Petition of William Johnson Gent. & c^a: to have the old Bridge Called Lewis's bridge Over the river Ny, repaired and Built at the County

Charge, the Same Was Overruled, but Ordered that Bartholomew Wood and his gang, and William Bartlett and his gang Do forthwith assist and help Robert Hutcheson and his gang Mend and repair the Said Bridge --

1 December 1730 O.S., Page 9
On Motion of robert King Overseer of the road from the County bridge to the Church road, and the road from the Church road to the County Line, to be Discharged from the Same, is granted, And Ordered that Charles filks Pigg Do Serve as Overseer in his room, And that all the Tithables That Serve under the former Overseer Do Serve under the Said Pigg in Clearing and repairing the Said roads --

2 February 1730 O.S., Page 13
On Petition of John Taliaferro & ffrancis Conway Gent. in behalf of themselves & Divers others, for a road from John Christophers to a Point of the fork of the Robinson from thence up the ridge the Convenientst way to the foot of Neals Mountain, the same is granted, And Ordered that Benjamin Porter the overseer of the Mountain road Continue his Said road from the Wilderness run to the fork of the Robinson, that David Phillips be made Overseer to Continue the said road from the fork of the Robinson the Nighest and Convenientest Way to the foot of Neals Mountain and that the said Overseers With their gangs Do make and Clear the Same --

2 February 1730 O.S., Page 13
On Petition of Benjamin cave to have the road that Comes from black Walnut run a Cross the river to the Mountain Chappel Devided at the river, is granted, and Ordered that Anthony Head be Overseer thereof, And that M^r. Robert Beverlys Tythables and Robert Darings Do Serve under him to help Make And Clear the Said road --

2 February 1730 O.S., Page 13
Henry Downes is appointed Overseer of the road from Taliaferro's road to ffoxs Point bridge, in the room of Joseph Hawkins Who is removed out of that Precints, And Ordered that all the Tithables that Served under the former Overseer Do help the Said Downs Clear and keep in repair the Same --

2 February 1730 O.S., Page 14
On Petition of William Bartlett in behalt of himself & Severall others to have Liberty of a bridle Way to go over the ford P m^r: Walkers Quarter over the River Po, to Church &c^a: Now the County bridge P m^r. John Snells being Carry'd away P the Late great freshes, the Same is

granted & Ordered that they have Liberty of a Bridle Way Especially till the Said Bridge be rebuilt or repaired --

2 February 1730 O.S., Page 14
On Petition of John Cook in behalf of himself and Severall others to have the road that is Ordered to be Cleared from Pleasant run to the Mountain road to goe through the fork of Pamunkey, It is Ordered that Andrew Harrison, Charles Stevens and John Cook Do Veiw and Layoff the Nearest and best Way, And that the Overseer With his gang Directly Clear the Same--

2 February 1730 O.S., Page 15
On Petition of Joseph Williams to be Discharged from being Overseer of the road from Cow Land up to the old road is granted, and Ordered that Robert Coleman be Overseer of the said road in his room, and that the Tithables that Served under the said Williams Do Serve under the Said coleman to help Clear & keep in good repair the Said road --

2 February 1730 O.S., Page 15
On Information made to the Court P Sundry People, that the County bridge P m^r. John Snells, being Carried away P the Late great freshes and that they Cannot role Tobacco or Pass without going Sundry Miles about &c^a. It is Ordered and Henry Goodloe, Joseph Brock and William Johnson Gent. or any two of them are Desired to Veiw and See if the old bridge Can be repaired and fitted up, and if not the Most fitting and Convenients Place to build a New one, and to agree With any Workmen as they Shall think best to repair the old or build a new one at the Countys Charge, and make report of their Proceedings to the Next Court

2 February 1730 O.S., Page 15
On Application made to the court in behalfe of the overseers of the roads where East north East bridge and the bridge over the River Ny, commonly called Lewis bridge, that the owners of the land adjoyning refuses to lett them fall or make use of any timber or plank to repair the said Bridges, which are much out of repair, It is ordered, and Henry Goodloe, Joseph Brock & William Johnson Gent^n: or any one of them are desird to agree with & to bye plank or timber as shall be P him or them thought necessary to mend and repair ye said Bridges, and that the Same Shall be paid and allowed of P the County at the laying the next levey --

3 February 1730 O.S., Page 18
On Petition of H.... Willis Gent. about a road to his Mill in the fork of Rappahanock river .. former Order not being Complyed With, On further Consideration it is ..... red that John Ashley be made Overseer And that he Clear the said road A .....g to the Directions of Robert Slaughter, Robert Green and William ..... Gent. or any two of them, and that they have Liberty to return a .... them to the Next Court that they think the Convenients to Clear the said road --

3 February 1730 O.S., Page 19
On the Presentment of the grand Jury against John Gordon for not keeping his Part of the road in repair within two Months. Ordered that the Same be Dismist --

Ditto --against Robert Hutcheson for not Keeping Lewis bridge in good repair Within two Months, Ordered that the Same be Dismist --

2 March 1730 O.S., Page 20
On Petition of Philemon Cavenaugh to be Discharged from being Overseer of the Rappahanock road, is granted, and Ordered that James Williams Do Serve as Overseer in his room--

2 March 1730 O.S., Page 21
On Petition of Michael Clore and George Wood for themselves and the rest of the Germans, to be free & Exempted from all roads, Except the road from the Island into the Main road that goes to Germanna ferry, the Same is granted --

2 March 1730 O.S., Page 21
On Petition of John Cook & John Red in behalf of themselves & others for a road to be made from Pleasant run to go through the fork of Pamunkey River up the Main ridge to the Mountains, Likewise Shewing the ill Conveniency of the road already Laid of P order of Last Court, Desireing to be Exempted from the Same. Ordered that Joseph Hawkins Do Veiw and Layout the Nearest and Best Way and make report of his Proceedings to the Next Court --

3 March 1730 O.S., Page 23
On Petition of M^r. Thomas Hill in behalf of himself and others the Inhabitants of ffredricksburgh Town in this County, to be free and Exempted from all County roads Except the Town, the Same is granted --

3 March 1730 O.S., Page 25
The Courts Order Appointing David Browne, Richard Blanton, William Bartlett and John Wiglesworth or any three of them to value the Timber Made use of in Building a Bridge over the River Po, in the Mine road, In Tobacco, the Said Order not being Complyed With, Joseph Brock & John Waller Gent. are Ordered & Desired to Value the Same, and Make report of their Proceedings to the next Court --

3 March 1730 O.S., Page 28
The Courts order appointing Henry Goodloe, Joseph Brock & William Johnson Gent. or any two of them to agree With Workmen Either to repair the old or Build a New Bridge over the River Po P M^r. John Snells, the Same Not being Complyed With, Ordered that the Same be Continued to the Next Court to compleat the Same --

3 March 1730 O.S., Page 29
On the Petition of Henry Willis Gent. about a road to his mill for Which Last Courts order appointed Robert Slaughter Robert Green & William Russell Gent. or any two of them to Layout the Nearest & Convenients, &c^a: and the Same being not returned & No further Proceedings therein, Ordered that the Same be Dismist --

3 March 1730 O.S., Page 30
M^r: William Russell is appointed to serve as overseer of the road from the Stones below wilderness bridge to Germanna ferry and to clear the old road P the flatt run Quarter that was formerly used, (In the room of Thomas ffarmer who is discharged) & ordered that all the male working tithables that served under the S^d ffarmer & Gordon the former overseer do help him clear the same --

6 April 1731 O.S., Page 31
Ordered that all Overseers of the roads be Continued, and that the Clerk Do Issue out Orders according to the Sundry Overseers for them to keep & Maintain their roads in good repair --

6 April 1731 O.S., Page 31
Robert Green Gent. is Discharged from being Overseer of the road from M^r. Bains Quarter to Maj^r. William Beverleys ford, And Anthony Scott is appointed to Serve in his room, and Ordered that all the Tithables that Served under the former Overseer Do Serve under the said Scott in Clearing & Keeping in good repair the Same

6 April 1731 O.S., Page 31
Ordered that the Clerk Do Issue Charles Piggs Order More fuller than before as he was appointed Overseer of the roads in the room of Robert King --

6 April 1731 O.S., Page 32
On the Petition of John Cook & c^a: for a road, Joseph Hawkins Who was Appointed and Ordered to Veiw & Layout the Same &c^a: failing to make a Sufficient report &c^a: Ordered that Andrew Harrison, Charles Stevens, Joseph Hawkins & John Cook Do Veiw and Layout the Nearest and best Way and Make report of their Proceedings to the Next Court --

6 April 1731 O.S., Page 33
Ordered that the Sheriff Do Wait on and Acquaint Col^o. Henry Willis to Compleat and finish the Wilderness run bridge (as he undertook) by the next Court, to Prevent his bond being Proseinted, on Complaint of said bridge Not being Done as P agreement made --

6 April 1731 O.S., Page 33
Ordered that Robert Hutcheson and his gang Do Clear the road from Lewis's bridge to Larkin Chews Mill --

4 May 1731 O.S., Page 34
On motion of M^r. Rice Curtis he having Liberty between this and the Last of September Next to pin down the Plank of the bridge over the River Po, by Mr. Snells, (it being now green) the Said Curtis keeping the said bridge Passable in The Meantime --

4 May 1731 O.S., Page 37 Grand Jury Presentment
We Do Present $Benj^a$: Porter for Not Clearing & Keeping in repair the road Between the Wilderness Bridge and the Mouth of the robinson River --

We Likewise present David Philips for not Clearing and Keeping in repair the road between the Mouth of the Robinson River and the foot of Neals Mountain --

We Likewise Present Thos. Cruthers for not Keeping the road in repair Between A run Called Arseformost, and the South Side River Po --

We Likewise present Robert Hutcheson for Not Keeping the Road in repair Between the County Line and the head of Greens branch --

We Likewise Present Colo. Henry Willis for Not building the Wilderness run bridge according to agreement --

We Likewise present Saml. Harris for Not repairing the Bridge over Middle River --

We Likewise Present the administrators of Capt. Larkin Chew Deceased for Not repair the Bridge over the River Po, the Bridge Built formerly by the said Larkin Chew--

We Likewise Present Bartholomew Wood for Not keeping the Road in repair between the Warf to the Bridge over the river Po By Information of John ffoster --

4 May 1731 O.S., Page 39
Joseph Brock and William Johnson Gent. Who was P the order of Court Impowered to agree With any Workman for the building the bridge over the river Po, by M^r. John Snells Mill Made return of the Agreement they made With M^r. Rice Curtis for building the Same, and his bond taken for Performance and Keeping the Same in good repair for the Space of Ten Years, Which Was Ordered to be recorded, And It is further Ordered that The Tobacco Agreed for Which is Six thousand Pounds of Tobacco, Cask & Conveniency, be Levyed P this County at the laying the Next Levy --

4 May 1731 O.S., Page 39
Joseph Brock and John Waller Gent. made return of this Court appointing them to value the Timber Made use of in Building a bridge over the river Po, in the Mine road &c^a. in Tobacco, Which is Viz^t. In Obeydiance to the Within Order We have veiwed the Mine bridge &c^a: and do value the Said Timbers of m^r. Thomas Turners, Made us of to three hundred and fifty Pounds of Tobacco -- the Same Was received, and Ordered that the said Tobacco by Levyed P this County at the Laying the Next Levy With Cost --

4 May 1731 O.S., Page 39
It is Agreed and Ordered that the Bridge over the River Ny, -- Commonly Called Lewis Bridge be rebuilt at the County Charge, Provided the Sum Do Not Exceed three thousand five hundred Pounds of Tobacco, for building the Same and Keeping it Ten Years in good repair, And Thomas Chew, Joseph Brock and William Johnson Gent. or any two of them are Desired and impowered to agree With anyone for the building the Same & that they make return of their Proceedings to the Next Court --

4 May 1731 O.S., Page 39
It is Ordered that Richard Blanton and his gang Do Put the bridge over Lewis's River in the Mine road forthwith in good repair --

5 May 1731 O.S., Page 46
Andrew Harrisson &^{c} failing to lay of the road petitiond for P John Cook &^{c}. & make report, the same is continued to the next court to be compleated --

5 May 1731 O.S., Page 46
Coll^o. Henry Willis not appearing according to summons to give his reasons why the wilderness bridge is not built & finished, the court being Satisfied that his absence from home, was the occasion the Said summons is continued to the next court for him to appear --

1 June 1731 O.S., Page 49
On Motion of Robert Beverley Gent. It is Ordered that Anthony Head and his Gang Do Clear the road from the Chappell road over the blue run to his Upper Quarter, and that Benjamin Cave With his gang Do Clear from Poplar bridge to the Rappadan River, and keep the Same in good repair --

1 June 1731 O.S., Page 49
On Motion of ffrancis Thornton Gent. It is Ordered that the old road from the Mountain run to Beverleys ford be Disannunlled, and that the road Strike out from the upper Side of the Mountain run bridge the Most Convenientest Way between ffrancis Thornton & Lawrence Battailes Quartes to the Mountain Tracts, and that M^r: ffrancis Slaughter be overseer thereof. And it is further Ordered that All the Tithables that Served on the other road help him Clear & keep in good repair the Same --

1 June 1731 O.S., Page 51
On Motion of John Scott Gent. he is Discharged from being Overseer of the road, from ffoxs Point to Hanover Line, and Ordered that James Barber Gent. Do Serve in his room, And that all the Tithables that Served under the former Overseer Do help the Said Barber to Clear & Keep in good repair the Same --

6 July 1731 O.S., Page 53
On Motion of Benjamin Porter Overseer of a road from the Wilderness bridge to the Robinson to be Discharged therefrom and Likewise to have

the said road devided into two Precints, the Same is granted, And it is Ordered that Dungan Bohannon be Overseer from the Wilderness bridge to the Pine Stake and that all the Tithables below the Catamount Do Serve under him, And that John Christopher be overseer from the Pine Stake to the Robinson, And all the Tithables that are above the Catamount Do Serve under him in Clearing & keeping in repair the Same --

6 July 1731 O.S., Page 56
On Motion of Charles ffilk Pigg he is Discharged from being Overseer of the roads from the County bridge to the Church road, and the road from the Church road to the County Line and William Hensley is Ordered and appointed to Serve in his room and that all the Tithables Which Served under the former Overseer Do help the Said Hensley Clear and Keep in repair the Said roads --

5 October 1731 O.S., Page 70
On Petition of William Beverley Gent. to have the old road Continued & his People Discharged from the New road &c^a: as in the Said Petition Mentioned, the Same is rejected, P reason it is the Oppinion of the Court that the New road Sets forth in the aforesaid Petition is the Most Convenient --

5 October 1731 O.S., Page 70
On Motion of John Mercer to have a Road Cleared from the Rappahanock road to M^{rs}: Jael Johnsons ferry Landing, is granted, And Ordered that James Williams the Overseer of the Main road aforesaid, & his gang Clear & Keep in good repair the Same --

5 October 1731 O.S., Page 73
David Phillips Appearing When Called to answer the grand Jurys Presentment for not Clearing & Keeping in repair the road Between the Mouth of the Robinson and the foot of Neals Mountain, the Court hearing the said Phillips Defense, Which was not Sufficient, Ordered that the Presentment be Dismist, the said Phillips Paying Costs --

5 October 1731 O.S., Page 73
Thomas Cruthers appearing When Called to answer the grand Jurys Presentment for Not Clearing & Keeping in repair the road Between Arseforemost & the South Side the river Po, His Defence being thought in Sufficient, Ordered that he be fined fifteen Shillings Current money With Costs --

5 October 1731 O.S., Page 73
Benjamin Porter Appearing When Called to answer the grand Jurys Presentment for Not Clearing & keeping in repair the road between Wilderness bridge & the Mouth of the Robinson River, the Court having heard the said Porter's Defence Which Was thought Sufficient -- Ordered that the said Presentment be Dismist ye s^d. Porter Paying Costs --

5 October 1731 O.S., Page 74
Robert Hutcheson Appearing When Called to answer the Grand Jurys Presentment for not Keeping the road in repair Between the County Line and the head of [missing] branch the Court having heard the Defence of the said Hutcheson & thought it is Sufficient, therefore Ordered that the Said Presentment be Dismist the said Hutcheson Paying Costs. --

5 October 1731 O.S., Page 74
The Grand Jurys Presentment against Col^o. Henry Willis for Not building the Wilderness bridge, the same is Ordered to be Dismist --

5 October 1731 O.S., Page 74
Samuel Harris Appearing When Called to answer the grand Jurys Presentment for Not repairing the bridge over the Middle river, the Court having heard the s^d. Harris Excuse, and thought the Same Sufficient, therefore ordered that ye Presentm^t. be Dismist, the said Harris Paying Costs --

5 October 1731 O.S., Page 74
The Grand Jurys Presentment against Thomas Chew & Larkin Chew adm^{rs}: c^a: Larkin Chew deced: for Not repairing the Bridge over the River Po, the Bridge built formerly by Capt. Larkin Chew Deced: the same is Ordered to be Dismist And It is Ordered that the bond Passed P Larkin Chew Deced: to Keep the bridge in repair &c^a: be Sent to the Kings attorney to be Put in Prosecution, but on the reading the Orders Larkin Chew one of the adm^{rs}: Afores^d. Came & assumed to build this Said bridge before the Last of December Next, therefore it is further Ordered that the bond Be Staidd from being Prosecuted --

5 October 1731 O.S., Page 74
Bartholomew Wood Not appearing according to Summons to answer the Grand Jurys Presentment for Not Keeping the road between the Wharf & the Bridge over the river Po. in repair. It is ordered that he be attached to answer the Same --

5 October 1731 O.S., Page 78
On Motion of Daniel Brown he is discharged from being Overseer of the road from the Church on Ta river to Coll°: John Wallers bridge, and George Carter is Ordered to be overseer in his room, and all the male tithables y^t Served under the Said Brown, do Serve under the Said Carter to clear & keep in repair the Said road and bridge --

7 October 1731 O.S., Page 79
Joseph Brock & William Johnson Gent^n: Justices, who were appointed P the Court to agree with workmen to rebuild the bridge over the river Ny commonly called Lewis Bridge, made their return of ye agreement made with Rice Curtis for two thousand pounds of tobacco with Cask & Conveniency to build the same & keep it in repair for the space of ten years & the Said Rices bond to perform the Same which was ordered to be lodged in the Clerk's office --

7 October 1731 O.S., Page 83
On Petition of John Cook & John Red in behalf of themselves & Several others for a road &c^a: Andrew Harrison &c^a: Who Was Ordered & appointed to Veiw & Layout the same failing to Make their return, It is Ordered that Robert Biggers & John Henderson Do Veiw and Layout the Nearest & best Way and make report of their Proceedings to the Next Court --

2 November 1731 O.S., Page 90
Ordered that John Grayson & Richard Blanton With their gangs Do forth With Put the bridge over the River Ny in the Mine road in repair --

2 November 1731 O.S., Page 92 County Levy

To Mr. Rice Curtis for building a bridge over y^e. River Po. as P bond	6,000
To Ditto for one over the river Ny.. as P bond	2,000
To M^r.Thomas Turner for Timber to build the Mine bridge &c^a...	618
To Mr Thomas Graves for repairing East North East bridge	400

2 November 1731 O.S., Page 94   Grand Jury Presentments
We Present And^r. Harrison for Not Clearing the rode he Was Appoint overseer Over according to the Order of Court for this Six Months Last Past –

We also Present John Durrett for Not Keeping the rode in repair Which he is Surveyor of for this Six Months Last Past –

3 November 1731 O.S., Page 95
Bartholomew Wood appeating When Called to answer the Grand Jurys Presentment for Not Keeping the road in repair between the Wharf to the bridge over the river Po. and he having Made his Same be Dismist, the Said Wood Paying Costs –

3 November 1731 O.S., Page 95
On Petition of Bartholomew Wood Overseer of the road from the County bridge over the River Po by Robert Kings to Nassaponax Wharf to be Discharged therefrom, the Same is granted, and Ordered that William Hust Do Serve as Overseer in his room, and that the Tithables Which Served under the former Overseer Do help the said Hust to Clear & Keep in good repair the said road –

3 November 1731 O.S., Page 97
On Petition of Thomas Witherby to have the road from the falls to the Wilderness Bridge be devided, (it being too long) is granted & ordered that the Said Witherby be overseer of the upper part Viz^t: from ye wilderness bridge to Coll^o: Alexander Spotswoods waggon road & that John Venton be overseer from the S^d waggon road to the falls, and that all the male tithables that used to Serve on the said road be equally devided between them, (according to the conveniency of the people as thay live by the S^d road) & ordered to help them clear & keep in repair the Said Road

3 November 1731 O.S., Page 97
Robert Biggers & John Henderson Made returne of their veiwing & laying of the road, as John Cook & John Redd & petitioned for, the Courts haveing taken the same into consideration are of Oppinion & Accordingly order that Joseph Hawkins Gent: do lay of the Said road Viz^t: from Terrys run up to the mountain road, [missing]ing above Mr Richard Bayleys land & houses, and that that road be deemed the County road & that John Cook be Overseer thereof, and all the Inhabitants above Terrys run do help him clear the Same –

1 February 1731 O.S., Page 99
On Petition of Ambrose Madison in behalf of himself & Severall others to have a road from the Dirt Bridge to ffredericksburgh Town Cleared and Shewing the ill Conviency of the old road Which is Now Cleared to the Said Town, It is Ordered that Ambrose Grayson Francis Thornton Junr. & John Grayson Junr. Gent. or any two of them Do Veiw both ways And Make report Which of the Same is Most Convenientst for the road to the said Town and return the Same to the Next Court –

1 February 1731 O.S., Page 99
On Petition of ffrancis Thornton Gent. to have a road Laid of and Cleared the Nearest and best Way from the beginning of the Mountain Tract by Jonas Ginkins's to the Inhabitants of the great Mountains, It is Ordered that Isaac Norman, John Hews, John Shaw and Joseph Bloodworth or any two of them Do Lay out and Mark the Most Conveniente Way and Make report of the Same to the Next Court –

1 February 1731 O.S., Page 99
On Petition of Ambrose Madison to have a road from the Camp run to Capt. Edward Ripons Quarter, is granted and Ordered that John Zachary the Overseer of the Same & that all the Tithables above the said Ripons Quarter Do Serve under the Said Zachary to Make Clear & keep in good repair the Said road –

1 February 1731 O.S., Page 99
On Petition of Henry Downes Overseer of the road from Crawfords tomb Stone to ffox Point to be Discharged from the Same is granted and Ordered that John Davis be Overseer of the road in his room, and that all the Tithables Which Served under the former Overseer Do help the Said Davis to keep in repair the said road –

2 February 1731 O.S., Page 100
On Petition of John Rucker to have a road from Benjamin Caves road to The upper End of Neals Mountain, it is Ordered that John Eddins, Michael Holt & Benjamin Cave or any two of them Do Layoff the Road the best & Most Conveniens Way & Make return of their Proceedings to the Next Court –

2 February 1731 O.S., Page 101
On Petition of Samuel Smith in behalf of himself & Severall others to have the road that M^r: Joseph Hawkins laid of from Terrys run to the

Mountain road Where John Cook Was appointed Overseer, and the Sundry fines remitted as the said Cook recovered against the Severall People refusing to Serve under him to Clear the Same, Disannulled. It is Considered & adjudged that the road Cleared by Andrew Harrison be Esteemed and remain the County road being the nearest that the Said Harrison Continue Overseer thereof, And it is further Ordered that ye road laid of P m^r Joseph Hawkins which John Cook Was Overseer of, be Disannulled, And that Part of ye s^d. petition about the fines be rejected–

2 February 1731 O.S., Page 105
Thomas Smith Gent is Appointed Overseer of the Rappahanock road in the room of James Williams Who is removed out of that Precinct, and Ordered that all the Tithables that Served under the said Williams Do Serve under the said Smith to Clear & Keep in repair the said road, And it is further Ordered that the Said Smith With his gang Do Make a bridge over the Hazle run and keep the same in good repair –

4 April 1732 O.S., Page 107
Samuell Ball Gent^n: is Discharged from being overseer of the road from Germanna to the mountain run bridge in the fork of Rappahanock river & Robert Slaughter Gentn: is ordered to serve as overseer in his room –

4 April 1732 O.S., Page 107
On motion of Thomas Smith Gent^n: who was appointed overseer of the Rapahannock road & to build a bridge over the Hazle run to have liberty to bye timber, or have some viewed & appraised, for the said use at ye Countys Charge, he being forewarned P the proprietors of the land adjoyning for Cutting or makeing use of any for that purpose was disabled in building the Said bridge, It is considered & Agreed to by the court & with the consent of James Williams proprietor of ye Land adjoyning to have John Grayson Junr: Uriah Garton, & James Roy or any two of them to view & appraise so much timber of the said James Williams as will be necessary & sufficient to build the said bridge & that ye S^d Williams be paid for the same at the Laying ye next County Levy, According to ye valuation to be returned to the next Court –

4 April 1732 O.S., Page 108
And it is likewise considered & Ordered P the Court that the old bridle way to M^{rs}: Jael Johnsons fferry landing be accounted ye county road to ye Said Ferry & to [indistinct] meeting house in ffredricksburgh And that the said M^{rs}: Johnson do [missing] gate or remove the fence, so that horses & Carts & may not be [missing] the Said Road to the Said

Inspecting house & Ferry, and that the Said Mr Thomas Smith Overseer do Let that road be kept in good repair with his gang –

4 April 1732 O.S., Page 108
On Motion of Henry Willis Gent^n: & P Causeys of Thomas Smith Gent^n: It is ordered that Thomas Hill, John Grayson & ffrancis Thornton Junr: or any two of them do Some time between this & the next court do view & lay ofe the most convenient way from ye S^d Willis plantation called Byrams plantation to the main road to Fredricksburgh and make report of their proceedings to the next court And it is further ordered that the Said Henry Willis Gent^n: have liberty to make use of the cart path from ye S^d Plantation by the Said Smiths to ye Said road (he puting up what fenceing he pulls down) In the meantime & both the said way be viewd & [indistinct]

4 April 1732 O.S., Page 108
On petition of John Bowman for a Road from Thorntons Quarter to Mitchels ford upon the river by Coll^o: Henry Willis Quarter up to the upper inhabitants in the little fork is granted & Ordered that the Said John Bowman be overseer thereof & that all the Inhabitants thereabouts do help him clear the same –

4 April 1732 O.S., Page 108
William Russell is discharged from being Overseer of the road from the corner Stones to Germanna ferry landing & William Bledsoe Gent^n: is ordered & appointed to Serve in his room & to goe as farr as the orchard of white oaks & have the liberty to clear the Said road the nearest way on the right hand from below his hous to the orchard of Oaks, And Ordered that Coll^o. Alexander Spotswoods four tithables, John Axford, Edward Teal, Richard Wrights Edward Abbot be added to his gang to help clear the same –

4 April 1732 O.S., Page 108
William Eddins is appointed Overseer from the Wilderness in the lower precincts of the old mountain road & Ordered to clear that part accordingly with his gang –

4 April 1732 O.S., Page 108
On Motion of Jerimiah Clowder Gentn: to have the road from George Musicks to the Mountain road (being too Large for one precinct) to be regulated and devided into three precincts for the better clearing the same is approved of & Accordingly Ordered that it be devided as

represented P the S^d Clowder & the persons recommend as Surveyors do Serve in ye Said precincts as followeth –vizt: –

Ordered that Thomas Pullam do clear the road [damaged or missing] from Coll^o: Aug^t: Moors Quarter to George Musicks, [damaged or missing] the Pond called the head of Pigeon, including the said Pond [damaged or missing] –

4 April 1732 O.S., Page 109
Ordered that Henry Lewis be made overseer of the road from the Pond, [d. or m.] head of Pigeon to Terrys Run and that Andrew Harrisson be [d. or m.] that part of the Road, and that the following hands do Serve under the said Lewis and help him clear the same –Viz^t: –

Mrs Mary Hawkins.	2	William Dyer …	1
John Wood ….	3	Thomas Allen …	3
Mr Winslows Q^u: …	4	George Musick ..	2
Mr Baylors Q^u: …	3	[d. or m.] Roberts –	1

4 April 1732 O.S., Page 109
Ordered that Andrew Harrisson be continued Overseer of the Road from Terrys Run to the Mountains road And that the following male Tithables belonging to the Severall people hereafter mentioned do help him clear the said –Viz^t:

John Henderson …	1	Mr. W^m ffleets Quarter	2
John Cook …	1	Coll^o: Aug^t: Moors.Q^r …	6
James Smith …	2	Mr Cha: Stevens …	3
Mr Taylors Quarter	4	Henry Chiles	1
Mr Chews Quarter ..	1	Thomas Cook	1
Samuel Hensley	2	John Cave	1
Capt: Jerr: Clowders Q^r ..	7	Mr Woodfolks Tennant –	2

4 April 1732 O.S., Page 109
Ordered that John Evans the overseer of the Roleing road from the ridge between Arseforemost and plentifull to Mattapony River in the room of

Thomas Crethers, and that the Male tithables belonging to the plantations hereafter mentioned do help him clear the same & keep the S^d road in good repair

M^r ffantleroys Quarter	3		Bushes Son in Law	1
George Dowdeys	1		Mrs Sanders Quarter	3
John Ward	1		Mr Garnetts Quarter	3
Robert Anderson	1		2: Quarters of	
David Brues	1		Coll° Corbins	6
Phillip Bush	1			

4 April 1732 O.S., Page 109
Ordered that Robert Cave (coll°: Corbins overseer) be overseer of the road from Mattapony River Side to the German road, & all the tithables in that precinct do help him clear the same –

4 April 1732 O.S., Page 109
On Petition of William Russell to have more tithables added to his gang as he was appointed overseer of is Rejected –

4 April 1732 O.S., Page 110
On Petition of Alexander Spotswood Esqr: for a Road to be laid of and cleared from his furnace door in this county into the main County road the best and nearest way is granted

5 April 1732 O.S., Page 112
Andrew Harrisson being called to answer the presentment of the grand Jury for not keeping the road in repair as he was Surveyor of, the court haveing heard what was said in his defence & Excuse, haveing considered the Same, ordered that ye presentment be dismist he paying the costs–

5 April 1732 O.S., Page 112
Ordered that John Evans and his gang do clear the road from the Mine road to Capt: Jerimiah Clowders roleing road the same way as ffranklyn cleared it and that John Durrett be discharged from being Overseer of the said road –

2 May 1732 O.S., Page 119
On Petition of ffrancis Slaughter to be Discharged from being Overseer of the road from the Mountain run bridge to the Mountain Track, is granted, and Ordered that John Roberts Do Serve in his room, And that

all the Tithables Which Served under the former Overseer Do help the said Roberts Clear & Keep in good repair the s^d road.

2 May 1732 O.S., Page 119   Grand Jury Presentments
We Present W^m. Hurt for not Keeping his road in good repair

We Likewise Present Abraham Bledsoe for Not Keeping his road in good repair –

We Likewise Present Jno: Davis for Not Keeping his road in good repair –

We Likewise Present Wm: Hinsley for not keeping his road in good repair –

2 May 1732 O.S., Page 120
The Last Courts order appointing Thomas Hill, John Grayson Junr & ffrancis Thornton Junr. or any two of them to Layoff & Veiw the Most Convenients Way from Henry Willis Gent Plantation Called Byrams Plantation by Thomas Smiths Gent. Plantation into ye main road (The Said Willis & Smith having Done the Same by Consent) and to Make report, have returned their Report Which is Viz^t. Spotsyvania In Obediance to the Order of the Worshipfulle Court Dated the 4th: of ap^{rl}. 1732 We the Subscribers in Company of m^r: Henry Willis & M^r. Thomas Smith did Veiw the road from the said Willis Plantation Called Byrams to the Main road, and the road Proposed by m^r.Smith And Do report that y^e. old road is Most Convenient. Thomas Hil, Jno: Grayson Junr. Fran: Thornton Junr Which report Was received & Ordered to be recorded –

2 May 1732 O.S., Page 121
On Petition of ffrancis Thornton Gent. to have a road Laid of & Cleared the Nearest & best Way from the begining of the Mountain Track by Johas Jenkins to the Inhabitants of the great Mountains, Last Courts order Appointing Veiwers to do the Same, being returned that it is Layd of & Mark, It is therefore Ordered that Isaac Norman be made overseer of the said road & that all the Tithables from M^r. Thorntons Upper Quarter to y^e. German ridge and So Down to the Mountain Track Do help him Clear & Keepin repair the Same –

6 June 1732 O.S., Page 126
On Petition of Henry Lewis Overseer of the road from the Pond, Called the head of Pigeon to Terrys run. to be Discharged, he Living ill Convenient & c^a. is granted. And Ordered that Thomas Allen be Overseer

of the said road in his room, and the following Tithables Which Served under the former Overseer Do help ye s^d. Allen to Clear the Same -- Viz^t. –

Mrs. Mary Hawkins –	8	W^m. Dyer ..	1
John Ward ...	3	Thomas Allen	3
Mrs. Winslows Quarter ..	4	George Musick	2
M^r.Baylors Quarter –	3	Joseph Roberts ...	1

6 June 1732 O.S., Page 126
Abraham Bledsoes Petition to be Discharged from being Overseer of the road from Crawfords Tomb Stone to the Mountain road, is granted, & Ordered that Richard Sharp be Overseer in his room, & that all the Tithables which Served under the former Overseer Do help the S^d. Sharp Clear the same –

6 June 1732 O.S., Page 126
On the Petition of Ambrose Madison in behalf of himself & others to have a road from the Dirt bridge to the ffredericksburgh Town Cleared, and showing the ill conveniency of the old road, Which was Now Cleared The Last Courts order for appointing Veiwers being Now returned that they have Veiwed the said roads and that the old road is the Most Convenients Way. therefore Ordered that they Same be Confirmed. & that the said Petition be Dismist –

6 June 1732 O.S., Page 127
On the Petition of John Rucker for a road from Benjamin Cave's road to The upper Neals Mountain, the Last Courts order appointing Veiwers to Layout the best & Convenients Way being returned that the best & Convenients Way for a road for the upper Inhabitants is to begin at the upper End of Francis Kirkleys Mountain and So Down the ridge between Bountifull run and the rappadan river into Caves road. therefore Ordered that Peter Rucker be overseer of the said road and that all the Inhabitants upon the South side of Thomas Smith's run Do help the said Rucker Clear the same road according to report –

1 August 1732 O.S., Page 129
Ordered that the road from the Chappell road to the Upper side of the Blew run Whereof Anthony Head is Overseer be Devided in two precincts and that the Said Head be Continued Overseer of the Upper Part of ye s^d. road and that from the Top of the hill above the blew run to Rippons Quarter, and that all the Tithables above the blew run to Rippons Quarter, and Robt. Dearing and the Tithables at Rippons Quarter Do help him Clear the Same and Ordered that Richard Winslow be appointed

Overseer from the hill on the upper Side of the blew run Just below Beverleys Mill to the Chappell road, And that all the Tithables hereafter Mentioned Viz^t. John Rucker & his Tithables, W^m. Crafords, John Andersons, Henry Downes, Thomas Jacksons, and Richard Winslows. Do help him Clear the Said road –

1 August 1732 O.S., Page 130
On Motion of Larkin Chew to have a road from the Mine road to the Church road, by Bush's tracts is granted. And Ordered that Abram Mayfeild be Overseer of the said road, and that all the Tithables at Mr John Walkers Quarter, Mr Paul Micou's Quarter, and John Durrets Do help him Clear the same –

2 August 1732 O.S., Page 135
On motion of Robert Riddle in behalfe of Mr Charles Chiswell & Company of the Iron mines, to have the bridge over the river Ny repaired (it being very much out of order) It is considered & Ordered that the same be repaired at ye County charge & Joseph Brock & John Chew Gent^n: are desired & Impowered to agree with some person to repair the same, who is to bring in his charge for the same at the laying ye County levy –

3 August 1732 O.S., Page 139
Ordered that George Purvis be Made Overseer of the road from Nassauponax to Snow Creek in the room of John Cammell and that all the Tithables between the said Places Do help him Clear & Keep in good repair the Said road –

3 August 1732 O.S., Page 139
Ordered that Samuel Harris Do Clear a road from his bridge to the Church road With the Same Tithables that Serve under him already –

3 August 1732 O.S., Page 142
William Hurt being Called toanswer the Presentment of the Grand Jury for not keeping his road in good repair, their being No such Man to be found Ordered that the same be Dismist –

    Abraham Bledsoe about Ditto. Acquitted –
    John D [destroyed] for Ditto the Same Ordered –
    William Hensley for Ditto the Same Ordered –

3 August 1732 O.S., Page 146
On the Petition of John Rucker to have all the Tithables that lie between the Robinson & Rappadan and above John Harsnipes, added to Peter Rucker's gang to Clear the road from the Upper End of francis Kirkleys Mountain and so Down the ridge between Beautifull Run and the Rappadan river into Caves road – is granted, and they are Ordered to help him Clear the Same, And Likewise It is Ordered that the road which David Phillips is Overseer of be Disannulled & Let fall and he is Discharged therefrom –

5 September 1732 O.S., Page 147
Joseph Brock Gent. is Ordered & Desired to take bond of Rice Curtis According to their Agreement for Building & Keeping in repair the Mine bridge over the River Ny Seven Years & c^a –

5 September 1732 O.S., Page 147
Ordered that all the Tithables on the River side from the falls to Nassauponax Do Serve under Mr Thomas Smith Overseer of the Road therefrom and Likewise It is further Ordered that the said Smith With his Said gang Do Clear & keep in good repair the road from M^r. Chiswells & c^a Mine Landing to Howes –

5 September 1732 O.S., Page 150
Ordered that William Tapp be Overseer of the road from Wharf to King and Queen County Road in the room of William Hust, and that all the Tithables that Served under the said Hust Do help the said Tapp to Clear the S^d. road & Keep the same in good repair, Except ffrancis Tunley & his Tithables Which are now Exempted –

6 September 1732 O.S., Page 152
On Petition of Philemon Cavenaugh in behalf of himself & others Setting forth the ill Conveniency of their Way already Cleared to ye Parish Church in the fork of Rappahanock and Petitioning for a More Conveniant Way to be Cleared from the fork road above Joseph Colemans into ye s^d. Place appointed for ye s^d. Parish Church. Ordered that Thomas Stanton & George Wheatley Do veiw the Ways & Make report Which is Most Convenients to the Next Court –

6 September 1732 O.S., P. 155
On Petition of Thomas Jackson & John Rucker in behalf of themselves and Others, to be Discharged from Working under M^r. Richard Winslow Overseer

of the Road from the Chapell to the Top of the Hill above Blew Water Run --Which Road Was granted by the Petition of Robert Beverley Esq. Last Court, and Shewing their being another road & c^a: the Said Petition is refferred to the Consideration of Next Court, and that Robert Beverley Esqr. have Notice given of their being Such a Petition Depending –

7 September 1732 O.S., Page 161
On Motion of John Taliaferro Gent. he is allowed Turn the road from Snow Creek to Mattapony Path With his own Tithables –

7 September 1732 O.S., Page 162
Francis Tunley is appointed Overseer of the Lower Part of Nassauponax Road in the Room of James Roy, and It is Ordered that the Tithables Which Served under the said Roy Do help the said Tunley Keep the said road in good repair.

7 September 1732 O.S., Page 162
Ordered that all the Tithables On the South side of the Falls road above Thomas Reeves be added & Do Serve under Charles Steward Overseer of the road.

3 October 1732 O.S., Page 163 County Levy
To James Williams for timber for building Hazle run bridge ...160

To Rice Curtis for new building ye mine bridge over the River Ny & keeping Same in repair Seven Years as P Bond. ...3436

3 October 1732 O.S., Page 166
Ordered that Richard [blank in book] be Overseer of the road from Nassawponax to the Upper Side of the Hazle run including the bridge and it is Ordered that all the Tithables on the river side between the said Places, Do Serve under the said [blank in book] to keep the said road in good repair –

3 October 1732 O.S., Page 166
Ordered that Thomas Reeves be Overseer of the road from Hazle run to the Upper Side of the falls run including the same, And the road out of the aforesaid road to the Inspectors Landing, And it is Ordered that all the Tithables On the river side, between the said Places Do Serve under the said Reeves to keep the said roads in good repair --

3 October 1732 O.S., Page 166
Ordered that Robert Hutcheson & his gang Do Clear the road from Lewis's Bridge to Chews Bridge, and Keep the same in good reapair –

3 October 1732 O.S., Page 166
On Petition of Thomas Pulliam to be Discharged from being Overseer of the road from Keys Mill Path to the Head [blank in book] is granted, And Ordered, that Joseph Thomas Do Serve in his room And that all the Tithables between the said Places Do Serve under the said Thomas to keep the said Road in good repair –

3 October 1732 O.S., Page 166
On Petition of Henry Chiles to have Liberty to Turn & alter the road that goes Over East North East bridge, Setting forth that the same is the intire hinderance of clearing his Plantation by Reason there is Not Sufficient Length & Breadth of Neither Side the Same & c^a: is granted and it is Ordered that the said Chiles have Liberty to alter & Turn the said road With the Approbation of John Waller & Edwin Hickman Gentlemen –

3 October 1732 O.S., Page 168
On Petition of William Hensley he is Discharged from being Overseer of the roads from the Mattapony Cleared to the County Line, and from the County bridge Into the said road and Ordered that Anthony ffoster be Overseer in his room and it is Ordered that the Tithables below the Church & On the South Side the River Po, to y^e. Rive Ta Do Serve under $y^e s^d$. ffoster in Clearing & Keeping the same in good repair –

7 November 1732 O.S., Page 169   Grand Jury Presentments
We Present Capt. William Bleadsoe for Not Keeping the road in repair from the Orchard of Oaks to Wilderness run –

* * *

We also Present Larkin Chew for Not Keeping the Bridge in repair Over the River Po -- .

7 November 1732 O.S., Page 172
On Motion of James Barber Gent. he is Discharged from being Overseer of the road from ffoxs Point to Hanover Line and Ordered that James Coward Do Serve as Overseer in his room and that all the Tithables which Served

under the former Overseer Do help the said Coward to Clear and Keep the same in good repair –

7 November 1732 O.S., Page 173
On the Petition of thomas Jackson and John Rucker for themselves and others, to be released and Discharged from the road Whereof M^r. Richard Winslow is Overseer from the Chappell to the Top of the Hill above blew Water Runk Setting forth that they serve on the road from ffox Point to Crawfords Stone Which is Ten or Eleven Miles Distance the same is granted, and Ordered that they be Discharged from Serveing under the said Winslow On the said road from the Chappell to the Top of Hill above blew Water run –

7 November 1732 O.S., Page 173
On Motion of John Evans he is Discharged from being Overseer of the roads from the ridge between Arse foremost and Plentifull to Mattapony river, and from the Mine road to Capt. Jerimiah Clowders roleing road the same Way as ffranklyn Clears it, and Thomas ffoster is appointed Overseer in his room And Ordered that the Male Tithables belonging to the Plantations hereafter Mentioned Do help him Clear the same and Keep said roads in good repair–

M^r.ffantleroys Quarter …	3.	David Bruce …	1.
George Dowdey …	1.	Philip bush …	1.
John Ward …	1.	Bushes Son-in Law …	1.
Robert Anderson …	1:		
	M^{rs}. Sanders Quarter …	3	
	Mr Garnetts Quarter …	3	
	2 Col^o. Corbins quarters …	6	

8 November 1732 O.S., Page 175
Ordered that William Phillips and William Stevenson's Tithables be added to John Davis gang, And it is further Ordered that the said Davis and his gang Do Clear the Road Down to Abraham Bledsoes instead of the Tombstone –

8 November 1732 O.S., Page 179
On the Petition of Philemon Cavenaugh in behalf of himself & others Setting forth the ill conveniency of their way allready cleared to the Parrish Church in the fork of Rapahanock & petitioning for a more convenient way to be cleared from the fork road above Joseph Colemans into ye Said place appointed for the Said Parrish Church the last order not being fully complyed with the Same is continued for the Said Thomas Stanton & George Wheatley the Gent^n. appointed P the Court to veiw the

ways and make a speciall & fuller report of the nearest & convenients way to the next court –

8 November 1732 O.S., Page 180
John Christopher is appointed Overseer of the road from the Pine Stake to Todds branch includeing the bridge over the same, and all the male tithables in his precincts below the said branch are Ordered to help him clear the Same & keep ye Said road & bridge in good Repair

8 November 1732 O.S., Page 180
Benjamin Porter is appointed Overseer of the Road from the Said Todds branch to the Robinson and that all the Male tithables that are above ye Said Branch are Ordered to help him clear the same

8 November 1732 O.S., Page 180
Ordered that George Pooles, Stephen Sharps, & Coll^o: Henry Willis Quarter (Where Downs lived) their male tithables be Added to William Tapps gang to help him clear the road as he is Overseer of –

8 November 1732 O.S., Page 181
Ordered that Ambrose Grayson gent^n: Do Value the timber that is made use of, for the building Nassauponax Bridge & make report of the same to the Court

6 February 1732 O.S., Page 181
The Honble: the General Courts order being Produced by Robt. Grame Gent about the old road from the Mountain Run to Beverleys fford to be again laid open and Cleared & repaired by the Tithables that Lives Most Convenient to the same and that the road from the upper side of the Mountain bridge Between ffrancis Thorntons & Lawrence Battailes Quarters to the Mountain Tract be Continued. It is Ordered by this Court that Anthony Scott be Overseer of the Said Old road from the Mountain Run to Beverleys Island ford And that the Tithables hereafter Mentioned Viz^t. Abraham Feilds Col^o _ Carters Tithables at Normans ford, Col^o. Beverleys, Capt. Greams, ffrancis Kirkley, William Paytons, Thomas Cook, James Ballard John Reed, George Roberts, Benjamin Roberts Andrew Glaspee, James Turner, George Smiths, Roger Oxford, John Haddox and George Sweeting, Do help the Said Overseer to make Clear & keep in good repair the said road –

6 February 1732 O.S., Page 181
On the Petition of the Hon[ble]. Alexander Spotswood Esq[r]. to have Liberty to alter and repair the road from the Wharf at Newport formerly Called Massauponax wharf to his Iron Mines at Tubal With his Own hands, it being Neglected by the Overseer of the Same, is granted, but it is Ordered he must not alter the same Without further Leave of the Court, But from Henry Martins to Mattapony Road –

6 February 1732 O.S., Page 181
On Petition of John Zachary to be released from being Overseer of the road from Camp run to Captain Ripons Quarter, he being removed out of that Precints, is granted, and James Stodghill is appointed Overseer in his room, and ordered that the Tithables above Ripons Quarter Do help him Clear and keep in repair the same –

6 February 1732 O.S., Page 182
On Petition of ffrancis Thornton Gent to have the road Continued and Carryed up from the Beaver Dam Swamp, to his Mill on Cannons River, is granted, And Ordered that Jonas Jenkins be Overseer of the same and that the said Thorntons Tithables and Jonathan Williams Do help the said Overseer to Clear & keep in Repair the said road, they being Excused & Exempted from all the other roads –

6 February 1732 O.S., Page 182
Ordered that all the Tithables from Daniel Brown up to John Ashleys be Added to Isaac Normans gang for helping to Clear the great Mountain road –

6 February 1732 O.S., Page 182
On the Petition of Thomas Graves Overseer of the road from East North East bridge to Keys Mill Path to be Discharged, is granted, and Ordered that Henry Chiles Do Serve in his room, and that all the Male Tithables Between the said Bridge and Path Do help the said Chiles Keep the said road in good repair –

6 February 1732 O.S., Page 182
William Tapp is Discharged from being overseer of the road from the Wharf to King and Queen County road, & George Poole is ordered & appointed to Serve in his room, & all the tithables that Served under the s[d]. Tapp are Order'd to Serve Under the S[d]. Poole to keep in repair the Same

6 February 1732 O.S., Page 183
On Motion of Robert Beverley Gent It is Ordered that the Tithables under M^r: Winslow Henry Downes George Anderson, William Crawford, William Crosswhite & John Rucker Do Clear the road from the Chappell run over the blew run to the said Beverleys Mill under the s^d: Winslow overseer of the said road –

6 February 1732 O.S., Page 183
On Motion of Larkin Chew to have the road altered from Capt. Jerimiah Clowders roling road to ye Mine road is granted, and Ordered that Thomas ffoster be overseer thereof With his Tithables Which are Viz^t. M^r. Fantleroys Quarter George Dowdey, John Ward, Robert Anderson, David Bruce, Philip Bush, Bushes Son in Law, M^{rs}. Sanders Quarter, M^r. Garnetts Quarter and two of Col^o. Corbins Quarters With Richard Jones to Clear & alter the said road By Bushes & franklyns the old way and keep the same in repair –

6 February 1732 O.S., Page 183
William Bledsoe Gent being Called to answer the Presentment of the Grand Jury for Not Keeping the road in repair from the Orchards of Oaks to the Wilderness run and he appearing and giving no reasonable excuse, Ordered that he be fined according to Law –

6 February 1732 O.S., Page 184
Larkin Chew being Called to answer the Presentment of the Grand Jury for Not keeping the Bridge in repair over the River Po. and he appearing and giving Sufficient Excuses, and the Court thinking the same Not Cognizable for them, Ordered that the same be Dismist –

7 February 1732 O.S., Page 188
On Motion of Goodrich Lightfoot Gent for a road to be Cleared from Stantons main road to the place where the Church is to be placed in the fork is granted and George Wheatley is appointed Overseer thereof –

7 February 1732 O.S., Page 188
On the petition of Philemon Cavenaugh for himself & others for a road to be Cleared from the fork road a little above Joseph Coleman, to the place appointed for the Church to be built, the Court haveing Considered the Veiwers report, are of oppinion & accordingly order that the petition be dismist –

6 March 1732 O.S., Page 192
On Petition of John Minor he is Discharged from being Overseer of the road, from Capt: Thomas Chews Mill to the County Line, And John Snow is appointed to Serve in his room, and Ordered that all the Tithables above Thomas Beals Quarter Do help the said Snow in Clearing and Keeping the said road in good repair –

6 March 1732 O.S., Page 194
On Petition of John Ashley and other Inhabitants for road from Col^o. Willis's Upper Quarter to Battle Mountain, the Same is granted, And Ordered that John Ashley be overseer thereof, and that all the Male Tithables above the said Willis's Quarter Do help him Clear & keep in good repair the said road.

6 March 1732 O.S., Page 194
Ordered that John Waller Junr. be overseer of the road from Wallers bridge to East North East bridge in the room of John Smith, and all the Male Working Tithables that Served under the said Smith, and all that have Lately Come into that Precinct Do Serve under the said Waller to help him Clear & keep in good repair the said road –

3 April 1733 O.S., Page 198
On Motion of John Taliaferro Gent for & in behalf of William Bledsoe Gent. overseer of the road from the Orchard of White Oaks to Germanna ferry Landing to have the Same Divided into two Precincts, from the said Germanna fferry to the Wilderness & from thence to ye said Orchard of White Oaks is granted, and Ordered that Richard Wright be overseer of the road from the said fferry to the Wilderness and that Edward Teal, John [blank in book], & Edward Abbott with the Tithables at Philemon Cavenaughs Plantation, Do Serve under the said Overseer to keep the said road in good repair, and it is further Ordered that William Bledsoe Gent be Continued overseer of the Other part of ye s^d. road And that the s^d. Tithables that Serve under Wright overseer be Discharged from Serving under Bledsoe on his road –

3 April 1733 O.S., Page 199
On Petition of Zachary Gibbs for himself and others to have a road from the Mountain road over [blank in book] above the Mouth of the Robinson river by M^r. Daggs Quarter to Elk Run is granted, and Ordered that those People Who is Convenients to & Wanted, the Same have Liberty to Clear it –

1 May 1733 O.S., Page 201
On Petition of Isaac Norman he is Discharged from being Overseer of the road from the Mountain Tract to Jonas Jenkins at the Great Mountain and John [blank in book] is Ordered to Serve in his room, and all the Tithables that Served under the said Norman Do Serve under the said [blank in book] to help Clear & Keep in repair the said road –

1 May 1733 O.S., Page 201
On Petition of Richard Wright overseer of the road from Germanna ferry to the Wilderness to have More Tithables added to his Gang, the same is Granted, And Ordered that all the Tithables that Live in the row below Brookes Run Do Serve under ye s^d. Wright in the said road –

1 May 1733 O.S., Page 201
On Petition of John Roberts he is Discharged from being Overseer of the road from the Mountain bridge to Mountain Tract, and Ordered that Charles Morgan Do Serve in his room, and that all the Tithables that Served under Roberts Do Serve under the said Morgan to keep the said road in good repair –

1 May 1733 O.S., Page 202   Grand Jury Presentments
We Par^t. John Oxford of the County aforesaid and Parish of S^t. George, for not Keeping his Part of the falls rode in repare – We Likewise Par^t. George Carter of the County & Parish afoursaid for Not not Keeping the Church Bridge and his Part of the rode in repare –

1 May 1733 O.S., Page 203
On Motion of John Chew Gent. for to have road Layd off and Cleared from the Bridge Called Lewis's to Chiswells & c^a: Iron Mine road, the Most Convenients Way, the Same is granted, and Ordered that George Home Do Lay out the same, and Make report of his Proceedings to the Next Court –

5 June 1733 O.S., Page 218
On Petition of Benjamin Winslow Gent for & in behalf of himself and Severall others for a road & c^a: according to the said Petition they have Liberty granted them to Clear the Most Convenients Way to the first Main road and Ordered that Lazarus Tilly be overseer thereof, and It is further Ordered that John Bush, Henry Martin, and Godfrey Ridge Do Veiw & Lay off the Same and Make report of their Proceedings to the Next Court –

5 June 1733 O.S., Page 219
George Carter being presented by the Grand Jury for Not Keeping his road & c^a. in repair, appeared according to Sumons and Made his Excuses and the Court being Satisfyed that the road Was Imediatly Cleared after the Storm as Soon as he knew of it, Ordered that the said presentments be Dismist Paying Costs –

3 July 1733 O.S., Page 221
Ordered that Abraham Bledsoe be Overseer of the road from Crawfords Tomb Stone to the Mountain road, In the room of Richard Sharp Who is removing out of that Precincts, and it is Ordered that The tithables Which Serve under the former Overseer help the said Bledsoe to keep the road in good repair –

3 July 1733 O.S., Page 223
On Petition of Lewis Davis Yancey for Liberty to alter and Turn the Mountain road Seting forth that it goes so Close to his Plantation that it is of very Great Prejudice to him, It is Ordered that ffrancis slaughter, John Roberts, David Mcmurrain or any two of them Do Veiw the Way Proposed by the said Petitioner and Whether the Said road may be Conveniently Turned Without any Prejudice to any Person and make report of their Proceedings to the next Court –

4 July 1733 O.S., Page 232
John Venton to answer the Presentment of the Grand Jury for not Clearing and keeping his road in good repair, the same Dismist he Paying Costs –

4 July 1733 O.S., Page 235
Benjamin Winslows Petition about a road is continued to the next court to finish the Same –

7 August 1733 O.S., Page 239
George Moore is appointed overseer of the road from the County Line to the head of Greens branch in the room of Robert Hutcheson who is removed out of this County, And Ordered that all the Male Tithables belonging to M^r: John Lewis The Honble. John Grymes, Esqr. M^r. John Chew, M^r: Larkin Chew, M^r. Richard Buckner and Those Tithables that Live at Capt. Thomas Chew's Old Plantation Do Serve under the said Overseer to Clear and Keep in repair the same –

7 August 1733 O.S., Page 240
On the Petition of Robert Eastham for a road for the use of ye Inhabitants above Muddy run in the fork of Rappahanock river to be Laid out and Cleared from Mitchels ford Crossing Muddy run the Most Convenients Way Down to the Mountain Road by M^r. ffrancis Slaughters, Ordered that M^r. ffrancis Slaughters and John Roberts Do veiw the Said Way and Make report of their Proceedings to the Next Court –

7 August 1733 O.S., Page 241
On the Petition of Lewis Davis Yancey to alter & Turn the County road by his Plantation, the same being refferred Last Court to ffrancis Slaughter, John Roberts & David M^{c}murrin to Veiw and report the same to this, Who having returned that the said Yancey, Might Turn the said road as Convenient another Way on his own Land & c^a: It is Ordered that the said Yancey have Liberty to Turn the Same With his own Tithables & Clear it accordingly –

7 August 1733 O.S., Page 241
John Nalle is Discharged from being Overseer of the road from the Mountain Tract to Jonas Jenkins at the Great Mountains and David M^{c}murrin is appointed to Serve in his room, and Ordered that the Tithables Which Served under nalle Serve under ye s^d. M^{c}murrain In keeping the said road in Good repair –

7 August 1733 O.S., Page 242
On Petition of Jonathan Gibson for himself and the rest of the Inhabitants of the Great Mountains for a road to be Cleared from Ruckers Road the Convenients Way to Offills Mountain, is Granted and Ordered that Thomas Jackson be overseer thereof, and that all the Male Tithables belonging to Anothony Thornton, Widow Conway, James Dyer & Benjamin Coward Do help the said overseer to Clear and Keep in Good repair the Said road they being Discharged from Ruckers road –

7 August 1733 O.S., Page 243
On Motion of Joseph Hawkins to have his Tithables Discharged from Working under Andrew Harrison on his road. And to be added to Lazarus Tillys road is Granted –

5 September 1733 O.S., Page 248
On Petition of Thomas Hill to be Discharged from being Surveyor of the High Ways, Streets & Publick Landings in ffredericksburgh Town, the Same

is Granted and Ordered that John Gordon Do Serve in his room, and that all the Tithables Which Served under the former Overseer Do Serve under the said Gordon to help keep the same in Good repair –

5 September 1733 O.S., Page 252
M^r: Benjamin Winslow failing to appear and prosecute his petition about a road from Rappahanock road to the mouth of Terrys Run &^{c}. Ordered that the Same be Dismist –

2 October 1733 O.S., Page 258
On motion of Joseph Woollfork in behalfe of himself & others to have the bridge called East North East bridge either built anew or repaired, it being very much out of repair, at the County charge is granted & Ordered that John Hollady & Joseph Thomas Gent^n: do agree with some workman to rebuild or repair the S^d bridge as thay shall think most proper & that ye Same be paid P the County at the laying ye next Levy –

2 October 1733 O.S., Page 259
Ordered that George Moore be Overseer of the road from Lewis bridge to Chew's bridge, and that the same Tithables Which Serve under ye s^d. Moore on the road from Greens branch to Lewis bridge, Do help him Clear & keep in Good repair the said road –

3 October 1733 O.S., Page 262
The Veiwers haveing made return of their report of laying of the Road as was petitioned for P Robert Eastham above muddy run in the fork of Rapphannock river to Mitchells ford, Viz^t: Wee have laid out and marked the best & Conveniens way wee can find is from Mitchells ford to Muddy Run above John Bomers where thay generally rowe over at, and from thence down between our plantations and to the main mountain road about the lower end of our Land, ffrancis Slaughter John Roberts; The same is confirmed P ye Court and ordered that John Bowmer be overseer thereof & that Henry Willis Gent^n: tithables be exempted from the said Road –

3 October 1733 O.S., Page 266
On Motion of m^r John Chew it is ordered that a road be cleared from m^r George Homes to Lewis bridge, and that Henry Rogers be Overseer thereof –

3 October 1733 O.S., Page 266
On Motion of Joseph Hawkins it is Ordered that John Fox be made overseer from The River Po to Germana road and Thomas ffoster be overseer from

the said River Po to the Pamunky road, and that both gangs do Joyn to make and repair the bridge Over the River Po, and that the tithables as lives convenient to each of the Said parts of ye Roads do Serve under the Said Overseers to clear & keep in repair the same –

6 November 1733 O.S., Page 266
On motion of Joseph Hawkings the following Tithables are ordered to serve under Thomas Foster to help him clear the road from the River Po to the Pamunky road and keep the same in repair, Viz^t: Colemans Quarter:3: ffantleroys Quarter:3: Coll^o: Corbins Quarter:3: Gortens Quarter:2: Robert Andress:1: David Bruce:1: Phillip Bush:1: Joseph Roberts:1: William [torn]:1: John Ward:1: George Musick:2: & m^r Baylors Quarter:4: and it is likewise ordered that John Fox and his gang have liberty to Cleare the most narest convenients way from the River Po to Rapahannock road

6 November 1733 O.S., Page 266
On motion of John Hollady Gent^n: in behalfe of himself and others to have liberty to clear a road from his mill road turning out by Thomas Sertains the most nearest and convenients way to Mattapony church, the same is granted and ordered that he be overseer thereof and that ye male tithables that Serve under him in his road do Serve under the Said Hollady to help him keep in repair & Clear the Said Church road –

6 November 1733 O.S., Page 267
On motion of Joseph Thomas Gent^n: to have the following hands Added to his gang to help him clear &keep in repair his road from Keys mill path to the head of Pidgeon is granted & Ordered that Coll^o: Augustine Moores Pattison Pullam, Thomas Pullam, James Rawlins, Robert Turner Anothony Golston & Henry Lewis & his own hands be added to his gang to clear ye S^d Road –

6 November 1733 O.S., Page 267   Grand Jury Presentments
Wee present the overseer of the road from Lewis bridge to the bridge over the River Po by Robert Kings of the parrish of S^t: George & County afforesaid within two months last past for not keeping the Road in repair –

Wee likewise present John Gorden of the parrish of S^t: George & County afforesaid for not keeping the Streets in the town of ffredricksburgh according to Law--

6 November 1733 O.S., Page 268
On Petition of Henry ffield to have a road cleared from the County road to his fferry the same is Rejected, P reason no fferry being yet kept in place appointed –

6 November 1733 O.S., Page 268
On Petition of Thomas Stanton to have Ruckers mountain road Devided into two precincts, the same is rejected –

7 November 1733 O.S., Page 269
On Peter Ruckers petition he is discharged from being overseer of the road from Benjamin Caves to ye upper end of ffrancis Kirkleys Mountain & ordered that John Garth do Serve in his room, and that all the tithables that did Serve under the Said Rucker do Serve under the Said John Garth to help him cleare & keep in repair the Said road –

5 February 1733 O.S., Page 270
On Petition of Augustine Smith Gent^n: for a road to be laid of from the end of the church road that comes into Stantons road to the said fferry that is appointed to be kept at Phillemon Cavenaughs fford & so through ye Said Cavenaughs' plantation into Coll^o: Alexander Spotswoods mine road is granted, & Ordered that Goodrich Lightfoot Junr and Lewis Davis Yancy do view & Lay of the same the most conveniens way & make report of ye Same to the next court –

5 February 1733 O.S., Page 270
On Petition of George Utz in behalfe of them Selfs & the rest of ye Germans to have the road that Michael Clore is overseer of devided, is refferred to the next court for consideration –

5 February 1733 O.S., Page 270
On Petition of George Carter to be discharged from being overseer of the Road from Coll^o: John Wallers Bridge to Mattapony Church (he haveing Served on the same about one year) is granted & Ordered that William Robinson Gent^n: do serve in his room & that all the male tithables that served under the said Carter do Serve under the Said Robinson to help him clear & keep the Said road in good repair –

5 February 1733 O.S., Page 272
On Petition of David M^c:Murrin he is discharged from being overseer of the Road from the mountain tract to Jonas Jenkins at the great

mountains, & Ordered that the said road be devided into two precincts and that Daniel Brown be overseer from Stonehous Mountain to Hughs, and Alexander M^c:Queen be Overseer from thence to Henry Winters and that Philemon Cavenaughs Male tithables be added to Alexander M^c:Queens gang with ye other tithables formerly granted, do help cleare the Same –

5 February 1733 O.S., Page 272
John Waller Junr: is discharged from being Overseer of the road from Wallers bridge to East north East Road, and John Wiglesworth is ordered & appointed to Serve in his room, and All the male tithables that belong to the said Waller do Serve under the Said Wiglesworth to help him clear & keep in good repair ye Said Road –

5 February 1733 O.S., Page 272
William Eddins & William Bohannon do Petition for liberty to have a road cleared from the fork of the Robinson to the head of beverleys mill run & from thence Straight to the main road is granted & Ordered that Zachary Gibbs be overseer thereof and that all the male tithables On the north side the butifull run do help the Said Gibbs clear the Same –

6 February 1733 O.S., Page 274
On Petition of George Home in behalf of himselfe and others for a Road from the west Side of the Mountains (which used to be through the pass of ye Mountains to Coll^o: Todds upper quarter which is now stopt up P ye S^d Todds clearing a plantation & fencing it in) that some persons may be appointed to view & lay out the most convenients way from the main road near m^r Colemans quarter by Coll^o: Todd's quarter where John Snow lives through the pass to the other, and thaough one Corner of ye Said plantation to the County line to ye s^d: Petitioners plantation the same is granted & ordered that John Rucker, John Davis & James Coward or any two of them do view and lay of the same & make report of their proceedings to the next court –

6 February 1733 O.S., Page 277
John Hollady & Joseph Thomas Gent^n: Appointed to agree with workmen for the repairing or rebuilding a bridge over East North East river, not haveing made returne, is continued to the next court to be compleated & returned –

7 February 1733 O.S., Page 280
George Moore being called to answer the present of the grand Jury for not keeping his road in repair, appeared & and made his defence, and is Excused P the court paving costs

7 February 1733 O.S., Page 281
John Gorden being called to answer the presentment of the grand Jury for not keeping the Streets in the town of ffredericksburgh according to Law the main Street being Judged to be in repair; he not Knowing ye other, is excused he paying costs –

5 March 1733 O.S., Page 285
John Rucker, James Coward and John Davison made returne of their View of the road petitioned for P George Home & ^{c}. from the west Side of the Mountains (which used to be through the pass of ye Mountains to Coll^o. Todds upper Quarter) Viz^t.In obedience to the within order wee y^e. Subscribers have mett and markt out y^e. most Convenient way from the Main Road near M^r. Colemans quarter up through y^e. pass of y^e. Mountains to Col^o. Todds upper quarter and not finding any way fitt to pass round have Markt a way through one Small Corner of y^e. said plantation & so along to y^e. County line near George Thomases quarter and find it may be made a good road for y^e. publick good, & part of y^e. Maine County road, witness our hands this first day of March 1733/4 _____ John Rucker, James Coward, John Davison, the Same is Confirmed according to y^e. Viewers report, And John Pied is appointed Overseer of the upper part & James Coward of the lower part of the said Roads, And Thomas Chew Gent is appointed & desired toddevide the said road in the hands, the Most Conveniens between the s^d. Pied & Coward to help them Clear & keep in repair the said roads

5 March 1733 O.S., Page 287
On petition of Augustine Smith Gent. for a road to be laid of to Cavenaughs fferry & ^{c}. the Viewers appointed, Not haveing returned their report, the Same is Continued to the next Court to Compleat the Same –

5 March 1733 O.S., Page 287
On Petition of William Wombwell in behalf of Charles Chiswell Esqr & company, belonging to y^e. Fredericks Ville, Ironworks, to have y^e. bridge over Nassauponax put in good repair, &, It is Ordered that Baldwin Colaugh be overseer of the said Road In the room of John Grayson Junr.; And that all the male tithables which Served under the said Grayson, do Serve under the said Collaron to help him Clear and keep in good repair the said road and bridge

5 March 1733 O.S., Page 287
John Holladay & Joseph Thomas Gent. returned y^e. bond & Agreement made with John Harris and John Wiggleswoath for the building &^{c}. of a bridge over East North East River, which s^d bond & Agreement is ordered to be lodged in the Clerks office –

6 March 1733 O.S., Page 291
On Petition of Jonathan Gibson setting forth that ye road from Ruckers road to y^e. olds mountains as Thomas Jackson was appointed Overseer of, that ye S^d Jackson for his own conveniency have cleared the Said road four miles out of the way, It is ordered that Benjamin Cave and Tandy Holeman do view and lay of the road the most convenients way & make report of their proceedings to the next court --

6 March 1733 O.S., Page 291
On petition of Thomas Stanton In behalf of himselfe & divers of the ffrontiers Inhabitants that ye road that goes P the name of Ruckers road as John Garth is appointed overseer may be continued up to Stantons River and that the said road may be devided into two precincts is granted and Ordered that John Garth be overseer of the upper part and David Phillips Overseer of the lower part, & m^r. Benjamin Cave is ordered & desired to devide the Said road & tithables y^t belongs or lives most convenients to each precinct between ye Said two Overseers according to his devizion, Each overseer with the said hands so devided are Ordered to clear & keep in repair the Said road

6 March 1733 O.S., Page 292
George Utz petition about haveing their German Road devided is Continued

6 March 1733 O.S., Page 295
On Petition of Richard Winslow to have hands added to his gang to help him make a bridge over blew run, is granted and ordered that Anthony Head with his gang do assist him to build the said bridge

2 April 1734 O.S., Page 297
On petition of William Todd Gent. to have a review of the Road that George Hooms petitioned for & was Confirmed last Court According to the Viewers report &^{c}. is granted, And Ordered that John Rucker, James Coward, John Davison & Benjamin Cave do View the Nighest way as Shall be Shown them P Each party, and report the Same Specially, And that the said Todd Do give the said Hoom's [indistinct] thereof

2 April 1734 O.S., Page 297
George Eastham is appointed Overseer of the road from Crawfords Tomb Stone to ffoxpoint, in the room of John Havis Deced, And ordered that all the tithables who Served Under the s^{d}d Davis do Serve Under the said Eastham to help him Clear & keep in good repair the said Road

2 April 1734 O.S., Page 297
On petition of Robert Slaughter Gent he is discharged of being Overseer of y^e. Road from Germanna to the Mountain Run bridge in the fork of Rappahanock river, And John Dowdey is appointed to Serve in his room & ordered that all the tithables which Serve Under the s^d. Slaughter, Do serve Under the s^d. Dowdey to help him Clear & keep in good repair the said Road

2 April 1734 O.S. Page 298
Lewis Davis Yancy & Goodrich Lightfoot Junr. made return of their view of the road, -- (petitioned for by Augustine Smith from the End of the Church Road that Comes into Stantons road to Philemon cavenaughs [missing] through the said Cavenaugh's plantation into Col^o. Spotswoods mine road) Viz^t. In obedience to the within order we have Viewed & laid out y^e. within mentioned Road according to y^e. order of Court, Lewis Davis Yancy, Goodrich Lightfoot, where upon the Court doo order that the said Road Shall Run and be as by the Said Viewers is laid of

2 April 1734 O.S., Page 298
On petition of George Utz & Michael Clore in behalf of themselfs & y^e. rest of the germans &^{c}. for to have their Road Devided &^{c} is rejected

7 May 1734 O.S., Page 306
On petition of Thomas Reeves to be Discharged from being overseer of the road from the Hazle run to the falls &^{c}. the Same is reffered to the next Court for Consideration about deviding the said roads & tithables that live adjacent &^{c}

7 May 1734 O.S., Page 307   Grand Jury Presentments
Wee present John Snow for not Clearing the road whereof he is Surveyor according to Law
...

We p^r. Sent James Sturgil for not keeping the road whereof he is Surveyor in Lawfull repair upon the Information of Henry Down's
We p^r. Sent the Surveyor of the road from Mountain run Bridge in the fork of Rappahannock river down to the said River at Germana for not keeping the s^d. road in Lawfull repair

7 May 1734 O.S., Page 310
The Viewers not haveing made report of their proceedings in Veiwing & laying of ye most Convenients way for a road out of Ruckers road to Offeilds Mountain as Jonathan Gibson Gent, petitioned for (& which as the said Gibson Complains that Thomas Jackson has Cleared it four Miles out of the way) The Same is Continued to be Compleated, & Ordered that Thos. Watts & Robt. Cave be added to Benj^a. Cave & Tandy Holeman (the two formerly appointed) And that they View both the ways & report the Same to the Next Court –

8 May 1734 O.S., Page 312
The Viewers not having made report of their proceedings in Viewing & laying of y^e. most Convenients way of the two roads now in dispute between William Todd and George Hoomes Gent. from the s^d. Hoomes Quarter through y^e. s^d. Todds upper Quarter at the Southwest Mountains, the same is Continued in order that they not only make report which is the best way, but also whither they find or Judge that both of them are Necessary for the publick good of the people to be kept open & Cleared

4 June 1734 O.S., Page 320
On petition of Anthony ffoster he is discharged from being Overseer of the Road from Mattapony Church to y^e. County line, and from Chew's bridge Into ye Church Road, And ordered that that part of y^e. Road from Hamm's bridge to the Church Road be added to the Same precinct, And that Thomas Graves be overseer thereof, And all the tithablas which Served Under the s^d. ffoster are ordered to Serve under the s^d. Graves to help Clear & keep in good repair the said Roads

4 June 1734 O.S., Page 320
Benjamin Cave, Thomas Watts, & Robert Cave, having made return of their proceedings in Viewing & laying out y^e. road in Dispute between Jonathan Gibson Gent. and Thomas Jackson, According to an Order of last Court Viz^t. In Obedience to an order of Court We the Subscribers have Viewed both ways And find the way that Tho Jackson has Cleared it is Considerable out of the way, We find the most Convenients way is to go the way that Jackson has Cleared it about a Mile and then Strke out and go by James Dyers, and over the foot of Neals Mountain And into offields track to the foot of offields Mountain Benj^a. Cave Thos. Watts, Robt. Cave, the same is Confirmed P the Court According to the Viewers report, And ordered that Thom^s. Jackson be overseer thereof, & that he with the hands formerly appointed him, no Clear And keep in good repair the S^d. road so Mark't out as aforesaid

4 June 1734 O.S., Page 321
On Motion of John Chew Gent. for to have the road drop't from Lewis's Bridge to Chew's Mill, & that he may have the liberty of Clearing the road from the River Po into the Church Road P the Race ground is granted

4 June 1734 O.S., Page 321
On Motion of Robert Slaughter Gent. for to have a road Cleared from the Glebe in the fork of Rappahannock River, to the South West Mountain Chappel is granted, in order that Christopher Zimmerman be overseer of that part of the S^d. road from the S^d. Glebe, to the Island ford, and all the Male tithables that live within four miles of the s^d. Road are ordered to help him Clear & keep in repair the Same, and that Thomas Jones be Overseer of that Part of s^d. road from the s^d. Island ford, the old way P W^m. Conico's P Maj^r. John Taliaferros Quarter Called Cattamount, And all the Male tithables y^t. live below y^e. s^d. Quarter (the quarter it Self Excepted) are ordered to Serve Under the S^d. Jones to help him Clear & keep in repair the Same, And that Alexander Waugh be overseer from the s^d. Cattamount Quarter to the South West Mountains Road, and all the Male tithables from the head of the Mountain Run Downwards, are ordered to Serve Under the s^d. Waugh to help him Clear & keep in repair the Same –

4 June 1734 O.S., Page 322
John Rucker, James Coward having made return of their proceedings in reviewing the roads In dispute between George Hoomes & W^m. Todd Gent^n. according to order of last Court Viz^t. In obedience to y^e. within order wee the Subscribers have meett & View'd both Rodes now in dispute between Coll^o. Todd & Geo^e. Hoomes y^e. one Shown us by John Snow Edward Franklin & others on y^e. upper Side of y^e. mountain's the other have Viewed again Shown us by Geo^e. Hoomes from the End of the main Rode by M^r.Collmans Quarter on y^e. Loar Side y^e. mountains Up through y^e. Pass to Coll^o. Todds Upper Quarter soe through one Corner of y^e. same to y^e. County line by y^e. s^d Hoomes Quarter, wee find to be much the best way & adjudge y^e. Same ought to be kept open as a Rode for y^e. publick, John Rucker James Coward, Ordered that the same be Confirmed according to the report of the Viewers, & it being a time that people are busied about their Crops, the Clearing of the s^d. road is refferred to the last of September Next Ensueing

4 June 1734 O.S., Page 323
John Dowdey being Called to answer the presentment of the grand Jury for not keeping his road in Lawfull repair, appeared & pleaded that he did not recieve the order of Court in time, which Excuse was P the Court

thought sufficient, It is therefore ordered that the Same be dismist, he paying y^e. Cost –

4 June 1734 O.S., Page 323
John Snow being Called to Answer Ditto, for & about Ditto, appeared & pleaded that the said road was in Dispute &^{c}. which Excuse was P the Court thought Sufficient, therefore ordered that y^e. Same be dismist, he paying y^e. Costs –

5 June 1734 O.S., Page 324
On petition of Thomas Reeves to be discharged from being Overseer of the Road from the Hazle run up to the falls, the Same is refferred till the list of tithables are return'd

2 July 1734 O.S., Page 329
On petition of Thomas Allen to be discharged from being Overseer of the road from the Pond Called the head of Pidgeon to Terry's Run is granted, and Lazarus Tilly is ordered to Serve in his room And all the tithables which Served Under y^e. s^d. allen are ordered to Serve under y^e. s^d. Tilly to help him Clear & keep in good repair y^e. s^d. Road

2 July 1734 O.S., Page 330
On petition of Charles Morgan to be Discharged from Serving as Overseer of y^e. road from the Mountain Run bridge to the Mountain tract, is granted And John Read is order'd & appointed to Serve in his room And all the tithables which Served Under the s^d. Morgan are Ordered to Serve under the s^d. Read to help him Clear & keep in repair the said Road

6 August 1734 O.S., Page 335
On petition of Edward ffranklyn in behalf of himself & Severall others, for to have a road from ffranklyns path out of y^e. James River Mountain Road ab^t. a Mile above Octonia Mill, the way y^e. s^d. ffranklyn Carried y^e. persons appointed to View the Road when Col^o. Todd & M^r. Hoomes were in dispute to the County line, Ordered that Benjamin Cave and Thomas Watts do View the Same, and make report of their oppinion & proceedings to the next Court

3 September 1734 O.S., Page 339
On petition of James Stodghill to be discharged from Serving as Overseer of the road from Camp Run to Cat^t. Ripons Quarter is granted; And Samuel Williams Is ordered & appointed to Serve in his room, And all the tithables which Served Under y^e. s^d. Stodghill are ordered to Serve

Under y^e. S^d. William's to help him Clear & keep in good repair y^e. said road

3 September 1734 O.S., Page 339
Ordered that Alexander Waugh be overseer of y^e. road from the Wilderness bridge to the pine Stake In the room of William Eddin's, and all the tithables which Served Under y^e. S^d. Eddings are ordered to Serve under y^e. s^d. Waugh, And that M^r. Conaway's q^r. John Christopher's & all the tithables below ye rockey run on y^e. back Side y^e. Southwest mountains be added to his gang to help him Clear & keep in good repair the Said road

3 September 1734 O.S., Page 339
Ordered the Beniamin Porter be overseer of y^e. road from the Pine Stake to y^e. point of the Robinson In y^e. room of John Christopher, and all the tithables which Served Under the S^d. Christopher are ordered to Serve Under y^e. s^d. Porter to help him Clear & keep in good repair the s^d. road

3 September 1734 O.S., Page 340
James Sturgil not appearing to Answer y^e. grand Jury's presentment for not keeping his road in repair &^{c}. the Court after hearing ye Excuses that were made in his behalf & Considered y^e. Same, Do order y^t. y^e. s^d. presentment be dismist Hee s^d. Sturgil paying Costs –

3 September 1734 O.S., Page 340
On petition of Thomas Reeves he is discharged from being Overseer of y^e. road from ye Hazle Run up to the falls, and James Williams is ordered and appointed to Serve in his room and all the tithables which Serve Under y^e. s^d. Reeves are ordered to Serve under the S^d. William's to help him Clear & keep in good repair y^e. s^d. road

3 September 1734 O.S., Page 343
The Overseer's Name being Omitted in the Order about the [damaged] Wm Todd & George Hoom's Gent were in dispute, It's Ordered that the [damaged] make y^e. s^d. Order more full according to the Courts order

4 September 1734 O.S., Page 344
Benjamin Cave & Thomas Watts having not made report of their oppinion & proceedings, in Viewing & laying out y^e. road y^t. that was last Court petitioned for P Edwd. ffranklyn y^t. the Same is Continued to y^e. next Court to be Compleated & returned

1 October 1734 O.S., Page 348
Ordered that George Purvis & the people which Serve Under him as Overseer of y^e. road from Nassauponax to Snow Creek, do Clear a road from the Lower Side of y^e. s^d. Nassauponax to y^e. County line P William Perry's, to meet a New road that is Cleared P order of Caroline Court

1 October 1734 O.S., Page 349
On petition of Lazarus Tilly overseer of y^e. road from the pond Called the head of Pidgeon to Terrys run for to have Some more tithables added to his gang out of Andrew Harrisons, is granted and Ordered that Capt. Clowders & Col^o. Moores tithables in the fork of Pamunkey, be added to help the s^d.Tilly Clear & keep in good repair the said road

1 October 1734 O.S., Page 349
On petition of Anthony Scott to be discharged from being Overseer of y^e. road from ye Island ford to y^e. road at y^e. Mountain run bridge, the same is rejected –

1 October 1734 O.S., Page 349
On petition of John Barnet in behalf of Col^o. John Grymes Esqr. for to have a road Cleared from the mountain road along M^r. James Taylors rowling road & from thence y^e. best & Nearest way to y^e. Rapidan river, is granted. And ordered that John Barnett be overseer therof, and that Mr. Thomas q^r. M^r. Mauldins plantation, Henry Kendol, sam^{ll}. Drake Zachary Gibbs, Thomas Walker, & Col^o. Grimes Male tithables are ordered to help him Clear & keep in good repair the Same

1 October 1734 O.S., Page 349
Ordered that George Moore be Continued overseer of the road as he was appointed P a former order of this Court, And that he with y^e. hands that Serves Under him, Clear & keep in good repair the Same

1 October 1734 O.S.,Page 349
On Motion of Charles Morgan in behalf of the Rev^d. M^r. John Beckett to have a bridle way from The Gleeb house to y^e. Church & ^{t}. Ordered that Robt. Slaughter Gent. James Kirk & James Pollard or any two of them do View and Lay out the most Convenients way & make report of their proceedings to ye next Court

1 October 1734 O.S., Page 351   County Levy
Tow William Eddins for to mend the Wilderness bridge …   300

To John Harris for building a bridge over East North East river as P
AgreemYt & ^{t} …   1400

2 October 1734 O.S., Page 355
On Motion made to the Court, It is ordered that the Clearing of the road that was in dispute between W^m. Todd & George Hoomes Gent, be put of till the last of this month –

2 October 1734 O.S., Page 356
Edward ffranklyn made return of Benjamin Cave & Thomas Watts Veiw of the road petition'd for P the said ffranklyn which is Viz^t. In Obedience to an Order of Court wee the Subscribers have Viewed the way that Edward ffranklyn Shewed us from ffranklyn's path above y^e. Octony Mill to y^e. County Line and wee find it to be a Very good way Benj^a. Cave, Thos. Watts, the S^d. Road is Confirmed According to y^e. Viewer's report, & Edward ffranklyn is ordered & appointed to Serve as Overseer & John Bryson, James Brown, Col^o. Todds Quarter that is Called Humphrey's Philip Bush & William Briant & their Male tithables are ordered to Serve Under the said ffranklyn to help him Clear & keep in good repair the Same

5 November 1734 O.S., Page 356   Grand Jury Presentments
Wee present the overseer of the road from y^e. Wilderness bridge to y^e. piny Stake for not keeping his road in Repair for these two months last past

…

Wee likewise present the overseer of the road from y^e. South west Mountain Chaple on Fox point Run for not keeping his road in Repair from the s^d. Chaple to Thomas Chews Mill Run for these two months last past

Wee likewise present Larkin Chew for not keeping the Bridge Called Chews bridge on y^e. River Po below Robert Kings for not keeping y^e. s^d. Bridge in Repare for these two months last past

5 November 1734 O.S., Page 357
On petition of ffrancis Taliaferro Gent^n. for to have the County road that goes by his house turned another Way through his land y^t., Ordered that James Williams Joseph Minter and ffrancis Tunley or any two of them do View the way Shown them P the said Taliaferro & make report of their proceedings to y^e. next Court with their oppinion thereof

5 November 1734 O.S., Page 358
On petition of John Hobson for to have Liberty to Clear a bridle way from his house into the Main Road that goes to Mattapony Church y^t. And to build a bridge over Robinsons run the Same is granted

5 November 1734 O.S., Page 360
James Kirk & James Pollard persons appointed to View & Lay of a bridle way from the Gleebe House in the fork of Rappahannock to the Church (petition for P y^e. Ref^d. Joh Bickett) made the following return Vis^t., At a Court held for Spotsylvania on Tusday October the first 1734, It is ordered on y^e. petition of Charles Morgin in Behalf of the rev'd M^r. John Becket to have a bridle way from y^e. Gleeb house to y^e. Church the most Convenient we y^e. Subscribers whose Names are heare Under writen have Begun at y^e. Gleeb and Marked out a road to y^e. Church According to y^e. Worship full Order y^e. Cort, James Kirk, James Pollard, which said View & return is received P the Court & ordered to be recorded

3 December 1734 O.S., Page 361
Ordered that John Landrum be appointed Overseer of the road from stone house Mountain to Hug's, in the room of Daniel Brown, and that all the male tithables which serve Under the said Brown, do Serve under y^e. said Landrum to help him Clear & keep in good repair the s^d. road

3 December 1734 O.S., Page 362
On petition of George Carter Setting forth that John Hikson having obtained an order of this Court for a bridle way through y^e. pet^{rs}. land, which said way (as it now goes) is very prejudiciall to him & praying that persons might be appointed to View & lay of The same, Ordered that John Durret, Thomas Hubbard & William Waller, or any two of them do View & lay of a bridle way the most Convenients, and the least prejudicial to the proprietor, & make return of their proceedings to the next Court

3 December 1734 O.S., Page 362
Joseph Minter & ffrancis Tunley persons appointed to View the way which ffrancis Taliaferro Gent petitioned for to turn the County road of &^{c}. made the following return Viz^t.In obedience to The within order we the Subscribers have View'd the way Shown us by Mr. Fran^s. Taliaferro and find it may be made as good a way as the road now goes & is as Convenient. Ordered that the Same be Confirmed & That the said Taliaferro have liberty to turn the said Road According to the plan Thereof lodged in Court (provided Col^o. Spotswood through whose land the same goes is willing Thereto.)

INDEX - SPOTSYLVANIA COUNTY ROAD ORDERS

Note: This index is arranged by subject: Personal Names; Bridges; Chapels, Churches, Glebes; Fords; Ferries; Houses; Mills; Geographic Features; Plantations; Rivers, Creeks, Swamps, etc.; Quarters; Signposts or Landmarks; Miscellaneous; and Roads

Personal Names:
Edward Abbot, 63, 76
Abraham Abney, 24, 40
Dennitt Abney Senr., 24
Dennitt Abney Junr., 18, 24
William Aginfield, 7
Thomas Allen, 64, 66, 67, 89
Conrad Ambergue, 34
George Anderson, 75
John Anderson, 68
Robert Anderson, 65, 72, 75
Robert Andrews (Andress), 25, 81
Thomas Appason, 8
ffrancis Arnold, 11, 12
John Ashby, 45
John Ashley, 52, 74, 76
John Askew, 13
Barnett Bain (Pain), 8, 13
John Bain, 13
Samuel Ball Gent:, 43, 62
James Ballard, 73
James Barber Gent., 56, 71
John Barnet, 91
William Bartlett, 6, 7, 10, 11^2, 32, 45, 46, 48, 50^2, 53
Lawrance Battaile, 56, 73
Mr: Mosley Battaley, 31
Mr Richard Bayley, 60
Mr: Robert Baylor, 24, 38
Mr Baylor, 64, 67, 81
Capt: Thomas Beal, 37, 38, 76
The Revd. Mr. John Beckett, 91, 93

Henry Berry, 18, 22, 31
Harry Beverley (mr, Capt:, Gentlm:), 8, 12, 13, 26, 42, 44
Robert Beverley (Esqr., Gent.), 50, 56, 69, 75
William Beverley Gentn:, 11, 12, 13, 15, 16, 17, 29, 57
Colo. Beverley, 73,
mr: Beverley, 22, 49
Stephen Bickham, 28
William Bickham, 28
Robert Biggers (Biggars), 59, 60
Edmund Birk, 28
John Blanton, 17, 21, 31
Richard Blanton, 27, 28, 31, 45, 46, 47, 48, 53, 55, 59
Abraham Bledsoe 37, 66, 67, 68, 72, 78
William Bledsoe (Mr., Capt:, Gentn:) 36, 40, 42, 63, 71, 75, 76^2
Joseph Bloodworth, 34, 61
Dungan Bohannon, 57
William Bohannon, 83
Patriack Bolding, 24
David Bolen, 7
John Bolen, 7
William Bolen, 7
John Bond, 36
Capt: Richard Booker, 11
mr Booker, 8
John Bowman (Bomer, Bowmer), 63, 80
Wm: Bradbourn, 24
William Brandegun, 112
William Briant, 92
Joseph Brock Gent., 11, 35, 41, 51^2, 53^2, 55^3, 59, 68, 69
Abraham Brown, 13, 15
Daniell Brown, 24^2, 27, 42, 48, 59, 74, 83, 93
David Browne, 53
James Brown, 92
David Bruce (Brues), 65, 72, 75, 81
John Bryson, 92
Mr Richard Buckner, 13, 25, 78
John Bush, 8, 12, 20, 21^2, 24, 25, 27, 28, 34^2, 35^2, 45, 77
Phillip Bush, 25, 27, 35, 65, 72, 75, 81, 92

Robert Bush, 25

Bushes Son in Law, 65, 72, 75

Walter Butler, 34, 35

Harry Cammell, 26

John Cammell, 11, 26, 68

James Cannon, 23

James Canny (Kenny), 5, 10

Alexander Carr 8, 12

Capt: Tho: Carr, 5

George Carter, 24, 59, 77, 78, 82, 93

Collo: Carter, 6, 73

Benjamin Cave 17, 23, 25, 27, 28, 36^2, 39^2, 44, 50, 56, 61, 67, 82, 85^3, 87^2, 89, 90, 92

David Cave, 43

John Cave, 64

Robert Cave (collo: Corbins overseer), 65, 87^2

Philemon Cavenaugh, 5^2, 9, 18, 30, 34, 36, 37, 39, 44, 52, 69, 72, 75, 76, 82, 83, 86

Richard Cheek, 27, 28, 33, 34

John Chew (Mr., Gentn:), 25, 68, 77, 78, 80, 88

Larkin Chew (Gentn:, Deced.), 6, 17, 18, 23, 24, 25, 26, 29

Capt: Larkin Chew, 6, 10, 12, 13^2, 15, 25, 27, 55

Mr: Larkin Chew, 58, 68, 71, 75^2, 78, 92

Thomas Chew Gent, 8, 13, 55, 58, 84

Capt: Thomas Chew, 16, 25, 30, 38, 76, 78

Mr Chew, 64

Henry Chiles, 64, 71, 74

Charles Chiswell Gent., 29, 30^2, 31, 32, 33, 35, 38, 39, 40, 41, 45, 47, 48, 68, 69, 84

John Christopher, 50, 57, 73, 90^2

Nicholas Christopher, 16, 19, 25, 29, 31, 37

Alexander Cleveland, 13, 28

Michael Clore, 22, 34, 52, 82, 86

Capt: Jerimiah Clowder, 12, 19, 20, 21^2, 23, 24, 30, 34^2, 35^2, 38^2, 40, 44, 45, 46, 63, 64^2, 65, 72, 75, 91

Frederick Cobler, 34

Michell Cock 13

Baldwin Colaugh (Collason), 84

Joseph Coleman, 69, 72

Robert Coleman 24, 51, 75

mr Coleman 83, 84, 88

Samuell Collins, 28
Thomas Collins, 28
ffrancis Conway Gent., 50
Widow Conway, 79
Mr.Conaway, 90
John Cook, 43, 51, 52, 53, 56, 59, 60, 62, 64
Thomas Cook 43,64, 73
Joseph Cooper, 34
Collo: Gawen Corbin, 14, 22, 36
Collo: Corbin 7, 10, 65, 72, 75
Benjamin Coward, 79
James Coward, 71, 83, 84, 85, 88
Wm: Craney (Crany) 8, 12
Wm. Crawford (Craford), 68, 75
Thomas Cruthers (Creders, Credders, Crethers), 25, 38, 46, 49, 54, 57, 65
William Crosswhite, 75
m^r Rice Curtis, 45, 46^3, 54, 55, 59^2, 69, 70
M^r.Daggs, 76
John Davison (Davis earlier), 61, 66, 72, 83, 84, 85^2
William Davis, 38
Robert Dearing (Daring), 50, 67
Joseph Delany, 26
Wm Dobbs, 24
George Dowdey, 65, 72, 75
John Dowdey, 86, 88
Thomas Dowdey, 23, 25
Thomas Downer, 29
Edward Downes, 12
Elias Downs, 28
Henry Downes, 50, 61, 68, 75, 86
samll. Drake, 91
John Durrett, 34, 35, 40, 60, 65, 68, 93
James Dyer, 79, 87
William Dyer, 64, 67
Robert Eastham, 79, 80, 85
John Eddins, 61
William Eddin(g)s, 5, 8, 9, 14, 19^3, 20, 26, 31, 33, 39, 40, 63, 83, 90, 92
Mr Edwards, 7

Lewis Elzey, 9

John Evans, 64, 65, 72

Mr ffantleroy, 65, 72, 75

Thomas Farmer 49^2, 53

Abraham Feild (ffeild), 7, 11, 12, 16, 17, 22, 73

mr Henry ffeild, 9, 37^2, 82

Mr John Finleson 12, 14^2, 18, 22, 23, 25, 45, 47

Richard fitzwilliams Esqr:, 30

Mr Wm ffleet, 64

Anthony Foster, 27, 28, 71, 87

John ffoster, 8, 13, 46, 55

Thomas Foster, 25, 27, 72, 75, 80, 81

John Fox, 80, 81

Joseph Fox, 34

Edward ffranklyn (Franklin), 6, 7, 8^2, 9, 11, 17, 20, 21, 33, 34, 35^2, 38^3, 40, 88, 89, 90, 92

Lawrence ffranklin, 8, 13, 14

Jno: Frauncis, 42

Thomas Gambal, 7

John Gambrell, 11, 12, 18

Thomas Gambrell, 12^2

mr: James Garnett, 25, 34

Mr Garnett, 27, 65, 72, 75

Uriah Garton, 62

John Garth, 82, 85

Zachary Gibbs, 76, 83, 91

Jonathan Gibson Gent., 79, 85, 87^2

Andrew Glaspee, 73

Anthony Goldson (Golston), 21, 24^2, 39, 44, 45, 81

Henry Goodloe Gentn: 13, 14, 15, 16, 23, 25, 26, 35, 51^2, 53

Mr: Goodlow, 8

John Gordon, 36, 39, 40^2, 49^3, 52, 53, 80, 81, 84

his Honr.The Governour, 46

John Grame Gentn:, 20, 27, 28, 36, 49, 73

Thomas Graves, 24, 59, 74, 87

Ambrose Grayson, Gentn:, 18, 26, 61, 73

John Grayson, 44, 59, 63

John Grayson Junr., 32, 61, 62, 66, 84

Capt. Greams, 73

Robert Green, 11, 12, 14, 28, 29, 34, 40, 42, 43, 52, 53^2
Collo: John Grymes Esqr., 26, 91
The Honoble: mr. John Grymes Esqr., 25, 78
Peter Gustavus, 31
John Haddox, 73
Samuell Ham, 8, 13, 14
William Hansford Gentn:, 27, 28
Capt: Hankings, 12
John Harris, 84, 92
Saml. Harris, 55, 58, 68
Andrew Harrison, 20, 21^2, 24, 27, 28, 30^2, 32, 49, 51, 54, 56, 59, 60, 62, 64^2, 65, 79, 91
John Harsnipes, 69
John Hawkings, 5, 8
Joseph Hawkins Gentn:, 16, 20, 29, 36, 50, 52, 54, 60, 61, 62, 79, 80
Mrs Mary Hawkins 64, 66
Nicholas Hawkins, 35, 44, 45, 46, 47
Anthony Head, 50, 56, 67, 85
John Henderson, 43, 59, 60, 64
Samuel Hensley, 33, 64
William Hensley, 57, 66, 68, 71
John Hews (Hughs), 61, 83
Edwin Hickman Gentn:, 12, 15, 16, 21, 23^2, 24, 30^2, 31, 36, 71
John Hikson, 93
Mr. Thomas Hill, 52, 63, 66, 79
John Hobson, 93
Phebe Hobson, 24
Tandy Holeman, 85, 87
John Holliday Gentn:, 12, 26 31, 41, 48, 80, 81, 83, 84
William Holloway, 28
Michael Holt, 18, 22, 35, 61
mr George Hoomes (Home, Hooms), 77, 80, 83, 84, 85, 87, 88, 89, 90, 92
Alexander Howard, 22
John Hows, 7
Thomas Hubbard, 8, 93
Wm. Hurt, 66, 68
William Hust, 60, 69
Robert Hutcheson, 46, 47, 49, 50, 52, 54, 55, 58, 71, 78
William Hutcheson, 11

Thomas Jackson, 16, 17, 19, 20, 68, 69, 72, 79, 85, 86, 87
George James, 18
Jonas Jenkins (Ginkins, Jenkings), 45, 61, 66, 74, 77, 82
Thomas Jermain (Germain), 20, 30, 31, 32
Widdow Jael Johnson, 17, 20, 21, 22, 31, 32, 41, 57, 62
William Johnson (mr, Gentn:), 25, 27, 28, 29, 35, 49, 51^2, 53, 55^2, 59
David Jones, 34
Richard Jones, 25, 34, 75
Thomas Jones, 88
Henry Kendal, 91
John Key, 12, 19, 21, 24, 27, 30, 32, 41, 44^2, 48
John Kimbrow (Kingbrow), 8^2, 9, 10, 12
Robert King, 7, 8^2, 13, 17, 25, 26, 27, 28, 29, 41, 48, 50, 54, 81, 92
James Kirk, 91, 93
Francis Kirkley, 11, 12, 18, 29, 34, 43, 47, 67, 69, 73, 82
John Landrum, 93
John Lawly, 12
John Lee, 19, 38
Henry Lewis, 64, 66, 81
Mr: John Lewis, 6, 25, 78
Zachary Lewis, 23, 24, 30, 31, 34, 36, 40, 45, 46
George Lightfoot, 9
Goodrich Lightfoot (Majr:, Gent.), 9, 36, 40, 42, 44, 75
Goodrich Lightfoot Junr, 82, 86
Mr: John Lightfoot, 34
Dennis Lindsey, 38
William Lindsey, 8
Dr: William Livingston, 20, 30
Wm: Lob, 12
James Mack Cullough, 38
William Mackconico, 23, 29, 38, 88
David Mcmurrain, 78, 79^2, 82
Alexander Mc:Queen, 83
Ambrose Madison, 11, 19, 61^2, 67
Henry Martin, 9, 14, 30, 42, 74, 77
Thomas Mauldin, 8
Mr. Mauldin, 91
Abram Mayfeild, 68

John Mercer, 43, 57
John Micou, 24
Mr Paul Micou, 68
John Minor (Mynor), 38, 76
Joseph Minter, 93, 94
Augustine Moore (mr, Collo:) 12^2, 35, 43, 64^2, 81
George Moore, 78, 80, 83, 91
Samuell Moore, 8, 10, 12
William Moore, 28, 31, 42, 43
Collo. Moore, 45, 91
Charles Morgan, 77, 89, 91, 93
John Mulkey, 9
George Musick, 30, 46, 49, 63, 64^2, 67, 81,
John Oxford (Axford), 63, 77
Roger Oxford, 29, 73
Collo: Man Page, 31, 32, 41
Thomas Parke, 36
William Payton, 48, 73
George Pemberton, 13
William Perry, 26, 91
David Phillips, 50, 54, 57, 69, 85
Thomas Phillips, 11, 19
William Phillips 72
John Pied, 84
Charles filks Pigg, 50, 54, 57
Edward Pigg, 13
John Pigg, 8
James Pollard, 91, 93
George Poole, 73, 74
Benjamin Porter, 29, 31, 37, 39, 47, 50, 54, 56, 58, 73, 90
Edward Price, 14
George Procter, 9, 20, 21
Daniel Pruett, 24
Paterson (Pattison) Pulliam, 45, 81
Thomas Pulliam, 44, 45, 64, 71, 81
George Purvis, 26, 68, 91
John Purvis, 8, 11, 26
John Quarles, 15

James Rawlins, 81
John Red(d), 52, 59, 60
John Reed (Read), 73, 89
Thomas Reeves 70^2, 86, 89, 90
William Rice, 13
William Richardson, 9, 15
Robert Riddle, 68
Godfrey Ridge, 77
Capt: Edward Ripon, 61, 74, 89
Benjamin Roberts, 73
George Roberts, 73
John Roberts, 28, 65, 77, 78, 79^2, 80
Joseph Roberts, 25, 38, 46, 67, 81
_____ Roberts, 64
Major Benja: Robinson, 11
John Robinson, 6, 10, 24
Collo: John Robinson, 13
William Robinson Gentn:, 82
Henry Rogers, 80
James Roy, 15, 62, 70
John Rucker, 20, 25, 29, 61, 67, 68, 69^2, 72, 75, 83, 84, 85, 88
Peter Rucker, 67, 69, 82
Peter Russell, 15, 16
William Russell Gentn:, 11, 17, 34, 36, 37^2, 42, 43, 53^2, 63, 65
Mr Nathaniell Sanders, 13, 25, 34
Mrs Sanders, 65, 72, 75
Anthony Scott, 53, 73, 91
John Scott Gent., 16, 38, 56
George Seaton, 31
John Sertain, 31, 81
Thomas Sertain, 12, 31
Richard Sharp 8, 14, 19^2, 20, 23, 38, 67, 78
Stephen Sharp, 17, 26, 73
_____ Skreen (Skrine), 6
Mr. ffrancis Slaughter, 48, 56, 65, 78, 79^2, 80
Robert Slaughter (Slatter) Gentn:, 17^2, 29, 33, 52, 53, 62, 86, 88, 91
Robert Slaughter Junr:, 9
Augustine Smith Gent., 6, 82, 84, 86

Maj: Augustine Smith, 12^2, 15, 49
George Smith, 36, 73
James Smith, 64
John Smith, 24, 36, 37, 76
Lawrance Smith, 17, 20
Samuel Smith, 61
Thomas Smith Gentn: 62^2, 63^2, 66, 67, 69
William Smith, 45
Capt: William Smith, 12, 21, 23, 25, 26
John Snell, 11, 23, 25, 26, 27, 28, 33^2, 34, 35^2, 50, 53, 54, 55
John Snow, 76, 83, 86, 88, 89
Henry Snyder, 13
James Sparkes, 14
Alexander Spotswood (Collo:, Esqr:), 6, 9, 14, 20, 27, 28, 36, 40, 44, 60, 63, 65, 74, 82, 86, 94
mr. Robert Spotswood, 22
Mr William Stanard, 13
Thomas Stanton, 9^2, 36, 69, 72, 82, 85
Abel Stears, 24
Charles Stephens (Stevens), 15, 17, 35, 43, 51, 54, 64
William Stevenson, 72
Charles Stuart (Steward), 14, 42, 70
James Stodghill (Sturgil), 74, 86, 89, 90
Robert Stublefield, 24
George Sweeting, 73
ffrancis Taliaferro Gentn., 93, 94
John Taliaferro (Mr., Gentn:) 5, 11, 15^2, 16, 19, 20, 26, 29, 50, 70, 76
John Taliaferro (Capt:, Majr.), 11, 19, 88
John Taliaferro Junr:, 25, 29, 37
Lawrance Taliaferro, 5
Mr: Robert Taliaferro, 16, 19
William Tapp, 17, 24, 25, 26, 33, 35, 37, 69, 73, 74
Mr. James Taylor, 91
Mr Zachary Taylor, 44
Mr Taylor, 64
Edward Teal, 63, 76
George Thomas, 84
Joseph Thomas Gentn:, 71, 80, 81, 83, 84
William Thomas, 7

Mr. Thomas, 91
William Thompson, 23
Anthony Thornton, 79
Francis Thornton (Mr, Gent.) 18, 19, 56, 61, 66, 73, 74
Francis Thornton Junr. Gent., 61, 63, 66
Samuel Tillary, 24
Lazarus Tilly, 77, 79, 89, 91
Phillip Todd, 16
William Todd Gent., 85, 87, 88, 90, 92
Collo: Todd, 83, 89, 92
ffrancis Tunley, 69, 70, 93, 94
James Turner, 73
John Turner, 28
Robert Turner, 12, 45, 81
mr. Thomas Turner, 55, 59
George Utz, 82, 85, 86
John Venton, 47, 60, 78
Mr John Walker, 68
Thomas Walker, 91
mr: Walker, 50
Jacob Wall, 6, 7, 11, 12
George Waller, 38
John Waller (Collo:, Gent.), 18, 24, 26, 31, 40, 53, 55, 59, 71, 82
John Waller Junr., 76, 83
William Waller, 18, 93
John Ward, 65, 67, 72, 75, 81
Thomas Warren 8, 13^2
William Warren, 8, 13
William Watters, 7, 10
Thomas Watts, 5, 8, 9, 87^2, 89, 90, 92
Alexander Waugh, 88, 90
George Wheatly, 9, 36, 37, 44, 69, 72, 75
Mark Wheeler, 8, 13^2
John Wigglesworth, 24, 40, 47, 48^3, 53, 83, 84
John Wilkings, 24, 37
James Williams, 30, 39, 43, 52, 57, 62^2, 70, 90, 93
Jonathan Williams, 74
Joseph Williams, 51

Samuel Williams, 89
Henry Willis Gentn., 22, 37, 44, 51, 53, 63, 66, 80
Henry Willis (Maj., Col.), 22, 23, 43^2, 45, 46, 47^2, 54, 55, 56, 58, 63, 73, 76
Benjamin Winslow (Mr., Gent.), 77, 78, 80
Mr. Richard Winslow, 67, 68, 69, 72, 85
Mr. Winslow, 64, 75
Mrs. Winslow, 67
Henry Winters, 83
Thomas Witherby, 60
William Wombwell, 84
Bartholomew Wood, 37, 50, 55, 58, 60^2
George Wood, 52
John Wood, 64
Mr Woodfolk, 64
Mr Woodfolks Tennant, 64
William Woodford, 20, 26
George Woodrofe, 40
Joseph Woollfork, 80
George Wooton, 29
Richard Wright, 63, 76, 77
Lewis Davis Yancey, 78, 79, 82, 86
John Zachery, Capt. Edward Ripons Overseer, 61, 74
Christopher Zimmerman, 34, 88
John D _____, 68
John _____, 76, 77
Joseph _____, 34
Richard _____, 70
William _____ Gent., 52
William _____, 81

Bridges
blew run bridge, 85
Capt: Larkin Chews bridge, 12
Chews Bridge, 70, 80, 87, 92
Church Bridge, 77
Corbins bridge, 22
County bridge, or simply bridge, 18, 34, 35, 37, 50^2, 57, 71
Dirt Bridge, 61, 67

East North east bridge, 11, 18, 21, 23, 24^2, 28, 37, 51, 59, 71, 74, 76, 80, 83^2, 84, 92

Fox point bridge, 29, 50

bridges on the Fredericksville Furnace Road or Mine Road, 33, 35, 38, 39, 41, 47, 48^2, 53, 55, 56, 59, 68, 69, 70

Hazel Run bridge, 39, 62, 70

Lewis Bridge on River Ny, 6, 27, 33, 34, 49^2, 51, 52, 54, 55, 59, 70, 77, 80^2, 81, 88

bridge over Lewis's River on the Mine Road, 56

Massaponax bridge, 15, 73, 84

Middle River bridge (Ta), 55, 58

Mine bridge, 40, 59, 69

Mine run bridge, 26, 31, 33, 39

Mountain bridge, 73

Mountain run bridge 34, 42, 56, 62, 65, 86^2, 89, 91

bridge over Mountain run in the fork of Rappahannock River, 14, 62

River Ny bridge, 47, 59

Poplar bridge, 56

Po river bridges, 55^2, 58^2, 59, 60^2, 71, 75, 81^2, 88

Po river bridge at Long Point above Abraham Brown's, 15

bridge over the River Po in the roling road Where Capt: Jerimiah Clowder is overseer, 34

bridge over the River Po att ffranklyns ford above Mattapony Church, 23, 25, 27, 28, 29

bridge over the River Po at John Snell's mill, 50^2, 51, 53, 54, 55^2

Todds branch bridge, 73

John Wallers bridge, 24, 37, 59, 76, 82, 83 bridge over

Warners River, 35

Wilderness Run Bridge, 5^2, 8^2, 9, 14, 29, 36, 40^2, 42, 47, 53, 54^2, 55, 56^2, 57^2, 58, 63, 75, 76, 77, 90, 92

Chapels, Churches, Glebes

the new chappell now a building 11^2

Fork Church, 69, 72, 75^2, 93

Glebe in the fork of Rappahannock, 88, 93

the new Church that is built on the river Ta, alias midle river 13^2, 14, 18, 33

Mattapony Church, 23, 24^2, 25, 26, 27, 34, 35, 38^2, 40, 45, 46, 71, 81, 82, 87, 93

Southwest Mountain Chappell on Fox point Run, 39, 50, 88, 92

Fords

Mr: William Beverleys fford, 29, 53

mr: Beverleys ford, 22, 56

Philemon Cavenaughs fford (ferry), 82, 84, 86

Devenports ford, 22
Elk river ford, 17
ffranklyns ford, 23
Island ford, 11, 88, 91
Mitchels ford, 63, 79, 80
Normans (normonds) ford, 28, 73

Ferries
Philemon cavenaughs ferry (fford), 82, 84, 86
Henry ffields ferry, 82
Germanna ferry, 13, 44, 52, 53, 63, 76, 77

Houses
Richard Bayleys land and houses, 60
mr Harry beverley's house, 13
Henry goodloes house, 13^2
John Hobsons house, 93
Mr Francis Thorntons house, 18

Mills
Beverleys Mill, 68, 75
Capt Larkin Chews mill, 27, 28, 54
Capt: Thomas Chews mill, 38, 76
Chew's Mill, 88
mill at or near the falls landing, 19
John Hollidays mill, 26, 31
Holidays mill, 24
Octonia Mill, 89, 92
John Snells mill, 55
Collo: Alexr: Spotswoods little mill, 9
Collo: Alexander Spotswoods old mill, 44
Francis Thornton's Mill on Cannons River, 74
Henry Willis's mill, 22, 23, 37, 43^2, 45, 47^3, 52, 53

Geographic Features
Battle Mountain, 76
Baylor's Mountain, 16
Blue Ridge (the Mountains or great Mountains), 16, 19, 43, 52, 61, 66, 77, 79, 82

little fork, 17, 22, 63
ffox point, 16, 17, 29, 50, 56, 61, 71, 72, 85, 92
German ridge, 66
the Island, 35, 52
James River Mountains, 89
Francis Kirkleys Mountain, 67, 69, 82
Little Mountains or South West Mountain, 5, 29, 36, 39^2, 44, 83, 84, 85, 87, 88, 90
Long Point on Po River above Abraham Browns, 15
Neals Mountain, 50, 54, 57, 61, 67, 87
Offills or Offields Mountain, 79, 87^2
ffork of Pamunkey, 35, 43, 51, 52, 91
fork of Rappahannock river, 14, 15, 22, 42, 43^2, 45, 47^2, 52, 62, 69, 72, 75^2, 79, 80, 86^2, 88, 93
Rich Neck, 7, 12^2, 21
Smiths Island, 13
Stonehouse Mountain, 83, 93
Todd's pass (probably), 83, 84

Plantations
Francis Arnold's plantation, 12
John Bains plantation, 13
William Beverley's plantation, 16, 17
Mr Richard Buckners plantations, 25
John Bush's plantation, 12, 24, 27
Byrams plantation, 63
Philemon cavenaughs Plantation, 76, 86
Capt: Thomas Chew's Old Plantation, 78
Capt. Jerimiah Clowders plantation, 12
Cow Land 7^2, 18, 31, 42, 43, 51
William Craney's plantation, 12
Cranwell's, 38
Thomas Creders plantation, 25
Edward Downes's plantation, 12
John Gambrell's plantation, 12
Thomas gambrell's plantation, 12
Edwin Hickman gentn:s plantation, 12
John Holliday's plantation, 12
John Keys plantation, 12
John Kimbrow's plantation, 12

John Lawly's plantation, 12
Mr: John Lightfoots Plantation, 34
Wm: Lobs plantation, 12
Ambrose Madison's plantation, 11
Mr. Mauldins plantation, 91
John Roberts's plantation, 80
Joseph Roberts plantation, 25
Thomas Sertain's plantation, 12
ffrancis Slaughter's plantation, 80
mr John Taliaferros plantation, 20
John Taliaferro Junr: plantation, 80
John Trusty's plantation, 12
Robert Turner's plantation, 12
Majr: Henry Willis plantation, 23
Henry Willis's plantation called Byrams near Fredericksburg, 63, 66

Rivers, Creeks, Swamps, etc.
Arnolds run, 22, 36
Arseforemost run, 21, 30, 49, 54, 57, 64, 72
Beautifull Run, 69, 83
Beaver Dam Swamp, 74
Beverleys mill run, 83
Blew, Blew Water, Blue Run, 56, 67, 68, 69, 70, 72, 75, 85
Bountifull run, 67
Brookes Run, 77
Camp run, 61, 74, 89
Cannons River, 74
Catamount, 57
Thomas Chews Mill Run, 92
Deep Run 15
Douglas, Duglass run 29, 30^2, 31, 40, 41
East North East Creek or River, 5, 8^2, 10, 25, 27, 83, 84, 92
Elk River, 17
Elk Run, 76
the falls 5^2, 9, 16, 30, 47, 60, 69, 70, 86, 89, 90
falls run, 70, 86, 88
fflatt run, 33, 49, 53
Glady fork, 34

Green's Branch, 6, 7^3, 9, 11, 24, 25, 33, 35, 44, 45, 46^2, 47, 55, 58, 78, 80

Hazell Run, 30, 31, 32^2, 34, 39, 41, 43, 62^2, 70^2, 86, 89, 90

Lewis's River on the Mine Road, 56

Massaponax 5, 11, 15^2, 68, 69, 70, 84, 91

Mattapony River 5, 8^2, 10, 12, 19, 21, 22, 30, 32^2, 49, 64, 65, 71, 72

Mine run, 26, 31, 33, 39

Mountain Run, 5, 6^2, 14^2, 15, 19, 29, 34, 42, 62, 73, 86^2, 89, 91

Muddy run, 22, 37, 79, 80

North anna River, 5, 7, 8, 20, 23, 30

Ny, Ney, Noy River 6^3, 9, 22, 32^3, 35, 39, 40^2, 41, 44, 47, 49, 51, 59^3, 68, 69, 70

Pamunkey Creek or River, 10, 19, 21, 25, 32^2, 35, 43, 51, 52, 91

the Pond called the head of Pigeon, 64^2, 66, 81, 89, 91

pleasant runn 19, 30, 49, 51

Plentifull, 21, 30, 49, 64, 72

Po River, 6^2, 15, 17^2, 23, 28, 29, 34, 35, 39, 41, 47, 48^2, 50, 53^2, 54^2, 55^3, 57, 59, 60, 71, 75, 80, 81^2, 92

Rappahannock or Rapahanack River, 14, 15, 22, 25, 28, 29, 30^2, 31, 32^2, 41, 42, 43^3, 44, 45, 47^2, 52, 62, 69, 72, 79, 80, 86^2, 88, 93

Rapidan, Rapid Anne, rapadan, Rappadan River, 13, 39, 56, 67, 69, 91

Robinson's River, 50, 54, 56, 57^3, 73, 76, 83, 90

Robinsons run, 93

rockey run, 37, 90

Thomas smiths run, 67

Snow Creek, 8, 11, 68, 70, 91

Stantons River, 85

Ta, Tay River, also Midle River, 13^2, 17, 18, 33, 55, 58, 68

Terrys run, 60, 61, 64^2, 80, 89, 91

Todds branch, 73^2

Walnut Branch on ye. North Side of Rappadan, 39, 50

Warners River, 33

White oak run, 35

Wilderness run, 5^2, 8^2, 9, 14, 29, 36, 40^2, 42, 47, 50, 53, 54^2, 55, 56^2, 57^2, 58, 63, 71, 75, 76, 77, 90, 92

Quarters

Bains Quarter, 29, 53

Lawrance Battailes Quarter, 56, 73

mr: Robert Baylors Quarter, 24, 38

Mr Baylors Quarter, 64, 67

Capt: Thomas Beals Quarter, 37, 38, 76
Robert Beverleys Upper Quarter, 56
William beverleys quarter, 11
Mr. Beverleys Quarter, 49
Alexander Spotswoods (Governors) quarter called the Bridge Quarter, 6^2, 7^2, 18, 22, 31, 42, 43
Mr Richard Buckners quarter, 13
Collo: Carters Quarter, 6^2, 12
Majr. John Taliaferros Quarter called Cattamount, 88
Tho: Carrs quarter, 12
Philemon Cavenaughs Quarter, 36
Mr Chews Quarter, 64
Capt: Jerr: Clowders Quarter, 64
Colemans Quarter, 81
mr Colemans quarter, 83, 84, 88
Mr.Conaway's qr., 90
Collo. Gawen Corbin's Stone hill quarter, 14, 36
2 Quarters of Collo Corbin, 65, 72, 75, 81
Mr. Daggs Quarter 76
Mr ffantleroys Quarter, 65, 72, 75, 81
fflatt run quarter, 33, 49, 53
Mr wm ffleets Quarter, 64
mr James Garnetts Quarter, 27, 34
Mr Garnetts Quarter, 65, 72, 75
Gortens Quarter, 81
Collo: John Grymes quarter, 26
Capt: Hankings Quarter, 12
Andrew Harrison's Quarter, 20, 21^2, 24
William Holloway's quarter 28
George Hoomes Quarter, 87
John Lewis's Esq. Quarter, 6
Mr Paul Micou's Quarter, 68
mr Augustine Moores lower quarter, 12, 35, 43
mr Augustine Moores upper quarter, 12, 35, 43
Collo: Augt: Moors Quarter, 64^2
Collo. Moores bridge Quarter, 45
Capt. Edward Ripons Quarter, 61, 67, 74
John Robinson Esq.'s Quarter, 6
Mr Nathaniell Sanders Quarter, 13, 34

Mrs Sanders Quarter, 65, 72, 75
mr. George Seatons quarter, 31
Skreen's, Skrein's quarter, 6^2, 12
Lawrance Smiths quarter, 17
Capt: William Smiths quarter, 12
Alexander Spotswood's Chestnutt Quarter, 14, 36
Capt: John Taliaferro's Quarter 11, 19^2
John Taliaferro Junr: Quarter, 29
Mr Taylors Quarter, 64
George Thomases quarter, 84
Mr.Thomas qr., 91
ffrancis Thorntons Quarter, 56, 73
Thorntons Quarter, 63
Mr. Thorntons Upper Quarter, 66
Collo: William Todds upper Quarter, 83, 84, 87, 88
Colo. Todds Quarter That is Called Humphrey's, 92
John Wallers quarter, 26, 31
Mr John Walkers Quarter, 68
mr: Walkers Quarter, 50
Collo. Henry Willis Quarter, 63
Collo: Henry Willis Quarter (Where Downs lived), 73
Colo. Willis's Upper Quarter, 76
Mr Winslows Quarter, 64
Mrs. Winslows Quarter, 67
William Woodfords Quarter, 20
mr: Woodfords quarter, 26
Woodfordes quarter, 17, 18

Signposts or Landmarks
Craffords, Crawfords TombStone, 37, 61, 67, 72, 78, 85
Pine Stake, 8, 11, 19, 20, 25, 57, 73, 90^2, 92
The Tombstone, 72
Stones marked AL below The Wilderness Bridge, 8, 9, 22, 23, 25, 27, 28, 39, 40^2, 49, 53, 63

Miscellaneous
Caroline County Court, 91
Charles Chiswell & Compa: Gentn:, also known as the Iron Mine Compa:, Fredericksville Furnace, etc. 29, 30^2, 31, 32, 33, 35, 38, 39, 40, 41, 45, 68, 69, 84

fall landing, 9

roleing house at the falls landing, 19

Fredericksburgh, 30^2, 31, 44, 52, 61, 63, 67, 79, 81, 84

meeting house in Fredericksburgh, 62

Germanna 5, 6, 7, 8^3, 9, 10, 13, 16, 20, 22, 23, 25, 27, 31, 36, 39, 40^2, 42^2, 43^3, 44, 49, 52, 53, 62, 63, 76, 77, 80, 86^2

Fredericksville Iron Works, 29, 30^2, 31, 32, 33, 40, 41, 42, 43, 84

The Courthouse (at Germanna), 36, 37

The fountain (at Germanna), 36

Germans, 13, 18, 52, 82, 86

Hanover County line, 16, 56, 71

Inspectors Landing, 70

Widdow Jael Johnsons fferry landing, 18, 20, 21, 22, 57, 62

King & Queen County, 67, 69, 74

Massaponax roleing house (James Kenny's), 6

Massaponax wharfe, 20, 24, 25, 37, 55, 58, 60^2, 69, 74^2

Mine Landing, 69

Mines, 27, 36

Newpost (Massaponax Wharf), 74

The Race ground, 88

Alexander Spotswood's furnace door, 65

Spotsylvania County line, 7, 12, 17, 23, 24, 35, 38, 44, 45, 46^2, 47, 50, 55, 57, 58, 71, 76, 78, 89, 91, 92

St. George Parish, 33^2

Tubal Iron Mines, 74

Roads

great road that goes up Arnold's run, 22

bridle path from Arnolds run to Germanna, 36

roleing road from ye ridge between Arseforemost & Plentifull to ye Southside of Mattapony river, 21, 30, 49, 51, 54, 57, 63, 64, 72

road from Baylor's Mountain to the Falls, 16

road to be continued past Beaver Dam Swamp to Francis Thorntons Mill on Cannons River, 74 road from mr Harry Beverley's house to the Church on the River Ta (alias Midle river), 13

road from the main road att mr: Beverleys ford in the fork of Rappahannock river up the little fork to Henry Willis's mill, 22

Bush's tracts, 68

road from Camp run to Capt. Edward Ripons Quarter, 61, 74, 89

Benjamin Caves road, 61, 67, 69

road from Benjamin Caves to ye upper end of ffrancis Kirkleys Mountain, 67, 69, 82

road from Benjamin Caves road to the upper end of Neals Mountain, 61, 67, 69, 82

Chappell road, 56, 67, 68, 70, 72, 75

road from the Chappell road over the blue run to Robert Beverleys Upper Quarter, and from Poplar bridge to The Rappadan River, 56, 67, 70, 72, 75

road from ye Church on the River Ta to ffranklyns road on the head of Greens branch, 33

Church road, 17, 28, 31, 50, 57, 68^2, 81, 82, 86, 87, 88

road from the end of the church road that comes into Stantons road to Philemon Cavenaughs fford (ferry), 82, 86

bridleway from the church road into Germanna road, 28

road to be cleared from Capt: Larkin Chews bridge up to Germanna road, 12

Road to Capt Larkin Chews mill by Lewis bridge, 27, 28, 54

road from John Christophers to a Point of the fork of the Robinson from thence up the ridge to the foot of Neals Mountain, 50, 54^2, 56, 57

road near Nicholas Christophers to extend up to the rockey run, 37

Jerimiah Clowders road, 24, 34, 35^2, 38^2, 40, 43, 45, 46, 65, 72, 75

Road from Capt: Jerimiah Clowders roleing road to mr Augustine Moores quarter in ye ffork of Pamunkey, 35, 43, 51, 52, 65

road from Capt. Jerimiah Clowders road to Mattapony Church over the bridge already built, 34, 35, 38^2, 40, 45, 46

Collo: Corbin's roleing road, 7, 22

road from Collo. Gawen Corbin's Stone hill quarter by William Eddins to the german road, 14

Collo: Corbin's upper road, 10

road from the county bridge to the Church road and roads thereabouts from the church road to the county line, 17, 50, 57, 71, 87

road from the County road to Henry ffields ferry, 82

road from Cow Land to the Bridge Quarter that comes into Germanna road, 7, 18, 31, 42, 43

road from Cow Land up to the old road, 51

road from Crawfords Tomb Stone to ffoxpoint, 85

road from Crawfords Tomb Stone to the Mountain road, 67, 78

bridle way from Devenports Ford to the great road that goes up Arnolds run and across the County to mrs: Jael Johnsons Ferry, 22, 62

road from the Dirt Bridge to ffredericksburgh Town, 61, 67

road from East North east bridge to the church on the river Ta (Mattapony Church), 18, 24, 28

road from East North East bridge to John Wallers bridge, 24, 37, 76, 83

road from East north East bridge to John Keys mill path, 24, 74

road from the County line to East North East bridge, 23, 40

road from the head of East north East to Mattapony Church, 25, 27

road from the mouth of East North East Creek to the head of Mattapony River towards Germanna, 5, 8^2, 10

East north East road, 83

road that goes to William Eddins below the orchard of White oakes, 9^2, 63

road that goes to the roleing house att the falls landing, 19

Falls road, 70, 77

road from the falls to Wilderness Bridge, thence to Collo: Alexander Spotswoods waggon road, Waggon road to falls, 5, 9, 47, 60

John ffinlesons road, 45, 47

gate and road alteration at fflatt run quarter, 33, 49, 53

ffranklyns path, 89, 92

road from ffranklyns path out of ye. James River Mountain Road abt. a Mile above Octonia Mill via the Todd Hoomes route, 89, 92

Franklyns road, 11^2, 20, 21, 33

Road from Franklyns road to the new chappell now a building thence to East North East bridge, 11^2

road from Fredericksville Iron Works to Rappahannock River nigh Fredericksburg, later called the Mine Road. Divided into segments: furnace, to ridge between Pamunkey & Mattapony, thence to the River Ny, thence to Hazel Run, 29, 30^2, 31, 32, 46, 47, 53, 55, 56, 59, 65, 72, 75, 77

road from the Mine road to the Church road by Bush's tracts, 68

road from the Mine road to Capt: Jerimiah Clowders roleing road, 65, 72, 75

Germans mountain road, 18

German Road, 9, 14, 65, 82, 85, 86

new road that goes from the German road up to the division, upper part and a lower part, 5, 9

old German road, 15, 17

new german road, 20, 21

road from the new german road at Franklyns road to mrs: Jael Johnsons Ferry, 20,21

bridle way from Germanna ferry into the road that comes by Collo: Alexander Spotswoods old mill, 44

Germanna Road, 6, 7, 9, 12, 17, 18, 20, 28, 31, 34, 42, 43, 80

Germanna old road, 7

road to be blazed and laid out from Germanna across the County to North anna River to go near Capt: Jerrimiah Clowders, 20, 23

road from Germanna along a Ridge to the Mountain Run and across to go between Collo: Carters & Skreens Quarter, 6, 12

road from the Glebe in the fork of Rappahannock River to the South West Mountain Chapel, divided into two sections: Glebe to the Island ford, and Island ford by Wm McConico's by Majr. John Taliaferros Quarter called Catamount, 88

bridleway from The Glebe to ye. Church, 91, 93

road to be laid out from Henry Goodloes house to the new Church that is built on the river ta, alias midle river through the lands of Barnett , Bain and Edward Pigg to Collo: John Robinsons rolling road, 13^2, 14

great road, 18

road from the great road to Widdow Jael Johnsons fferry landing, 18

county road to Greens branch, 17, 24

road from the County line to Greens branch, 24, 25, 35, 44, 45, 46^2, 47, 55, 58, 78

road from the head of Greens branch to Collo: Corbins roleing road, 7

road from the head of Greens branch to Germanna old road, 7

road from Green's branch to Lewis bridge, 80

road from the head of Greens branch crossing the River Ney to the Germanna road, 6, 9

road from Hazle run to the Upper Side of the falls run, and out of same to the Inspectors Landing, 70, 86, 89, 90

bridle way from John Hobsons house into the Main Road to Mattapony Church, 93

road from John Hollidays mill to the main road that goes by John Wallers above his quarter to the Church, 26, 31

road from John Holladays mill road by Thomas Sertains to Mattapony Church, 81

road from ye Island ford to ye. road at ye Mountain run bridge, 91

rolling road into the main road from the Island in the first fork of White oak run, 35, 52

road to be cleared from the end of the road whereof Jacob Wall is overseer down to the Island ford & to Beverley's quarter, 11

John Keys mill path, 24^2, 44, 45, 71, 74, 81

road from Keys Mill Path to the Head of Pidgeon, 71, 81

road from John Keys path that goes to Holidays mill, & so upwards to the road that Jeremiah Clowder is overseer of, 24, 44, 45

Road cleared by King & Queen Inhabitants, 6

King & Queen County road, 7, 69, 74

road from King & Queen County road to the lower branch of Green's branch, 7 road from the bridge on the river Noy below John Lewis esqrs quarter to his Honbl ye Governors quarter called the bridge Quarter, 6

road from mr George Homes to Lewis bridge, 80

road from Lewis bridge to Chew's bridge 80, 88

road from Lewis's Bridge to the Mine road, 77

road from Lewis bridge to the bridge over the River Po by Robert King's, 81

road Cleared from Mr: John Lightfoots Plantation into Germanna road, 34

road from the Little Mountains (South West Mountain) to Wilderness Bridge (later South West Mountain Road), 5, 29, 36, 39, 61

road from the west Side on the Southwest Mountain through Todd's pass, 83, 84, 85, 87, 88

road to be cleared on the south side of the little mountains to the Hanover line. Divided into segments: Hanover line to ffox point, and ffoxpoint to Taliaferro's road, 16, 20, 24, 29, 36, 50, 56, 71

road from the Pine Stake to Ambrose Madison's plantation, 11, 19, 25

road from Nassauponax to ye. County line by William Perrys to meet a New Road Cleared by order of Caroline Court, 91

road from Nassawponax to the Upper Side of the Hazle run, 70

Massaponax, Mausauponax, Nassauponax road, 14, 15^2, 17, 30, 31, 42, 70

roleing road from Nasauponax road to Mattapony main road, 17, 31

Massaponax new road, 15, 17

road from Massaponax to the falls, 5

road from (Massoponax) Wharf to King and Queen County Road, 69, 74 road from the county bridge to Massauponax wharfe, 17, 24, 25, 37, 55, 58, 60^2

road from the mines to Masauponax Mattapony main road, 5, 15, 17, 31, 74

rowling road from Mattapony main road to the massauponax road, 15

Mattapony Path 70

road from Mattapony River Side to the german road, 65

road from the bridges over Middle River (Ta) to the Church road, 68

great Mountain road, 74

Mountain road, 22, 38^2, 44, 47, 49, 50, 51, 60, 62, 63, 64, 67, 76^2, 79, 80, 91

road from the mountain road along Mr. James Taylor's rowling road to Rapidan River, 91

road from the Mountain road over _____ _____ above the Mouth of the Robinson river by Mr. Daggs Quarter to Elk Run, 76

road in the fork of Rappahannock river to run from Mitchels ford Crossing Muddy run Down to the Mountain Road by Mr. ffrancis Slaughters, 79, 80

old mountain road 63

road from the Mountain run to the end of Sharps path comeing into the roleing road, 19

mountain run road, 14

road from Germanna to the Mountains by way of Mountain Run in the fork of Rappahannock River, 14, 16, 17, 42, 62, 86^2

Road from the Mountain Run to Mr: William Beverleys fford, 29, 53, 56, 73

Road from Mountain Run to Bains Quarter, 29

Road from Bains Quarter to Mr: William Beverley's fford, 29

Mountain Tract(s) Track, 56, 61, 65, 66, 73, 77^2, 79, 82, 89

road from the beginning of the Mountain Tract by Jonas Ginking's (Jenkins's) to the Inhabitants of the great Mountains. Later divided in two: Stonehouse Mountain to Hughes, and Hughes to Henry Winters, 61, 66, 77, 79, 82, 83, 93

road from the Mountain bridge to the Mountain Tract, 77

road from the Mountain run bridge to the Mountain Track, 65, 89

new road from the Mountain run in the fork to the upper inhabitants, 15

road from Northanna to Fredericksburgh over the Hazell run, 30, 34

offields tract, 87

Pamunkey road, 81^2

road up Pamunkey to the head of Mattapony branches to be carried toward the mountains as far as pleasant runn, 19, 30

road from the Pond called head of Pigeon to Terrys Run, 64, 66, 89, 91

road from the pine Stake to ye. point of the Robinson, 90

road from the Pine Stake to Todds branch, 73

road from the Pine Stake to John Taliaferro's Quarter, 19, 20

road from the River Po to Germanna Road, 80

road from the River Po to the Pamunkey road, 81^2

road from the River Po into the Church Road by the Race ground, 88

new road to Parish Church in the fork of Rappahannock to be cleared from Joseph Colemans to church site, 69, 72, 75 fork road, 69, 72, 75

Rappahannock Road, 39, 44, 52, 57, 62^2, 80, 81

road from the Rappahannock road to Mrs: Jael Johnsons ferry landing, 57

road from Rappahannock road to the mouth of Terrys Run, 80

road from the ferry at Germana to Smiths Island up the river rapadan, 13

road from the dividing (county) - line to Rich Neck, 7, 12

road from Rich Neck upwards to head of Mattapony river, 12

road from Rich Neck upwards to the road that Jerimiah Clowder Gentn: is overseer of, 21

roleing road from John Roberts by Robert Greens to normonds ford on the north side of Rapahanock river, 28

road from John Robinsons Esqr Quarter over the river Po by Larkin Chews to Massaponax roleing house (James Kenny's), 6, 10

Collo: John Robinson's rolling road, 13

road from the fork of the Robinson to the head of beverleys mill run to the main road, 83

roleing road, 19

Ruckers mountain road, 82

Ruckers Road, 79, 85^2, 87

road from Ruckers Road to Offills Mountain, olds mountains, 79, 85, 87

Ruckers road extended to Stantons River, 85

Sharps path, 19

Majr: Augustine Smiths road, 12

road to be laid out, and cleared from Majr: Augustine Smiths road to the upper inhabitants, 12

road from Snow Creek to Massaponax, 8, 11, 68, 91

road from Snow Creek to Mattapony Path, 70

Road from Alexander Spotswoods Quarter called Bridge Quarter down the Ridge between the Rivers Po and Ny and over the same by Lewis Bridge to the road cleared by the King & Queen County inhabitants, 6, 7

road from Alexander Spotswood's furnace door into the main County road, 65

Road from Collo: Alexr: Spotswoods little mill to Thomas Stantons, 9

Colo. Spotswoods mine road, 86

Colo: Alexander Spotswoods waggon road, 60

Stantons road, 75, 82, 86

road from Stantons main road to the place where the Church is to be placed in the fork, 75

road from the stones marked AL below the Wilderness bridge to Germanna (ferry), 8, 12, 22, 23, 25, 27, 28, 39, 40^2, 49, 53, 63

road from the Stones to the road that goes to William Eddins below the orchard of white oakes, 9, 63, 71, 75, 76

from there down to the fall landing, 9, 63

Taliaferros road, 17, 29, 50

Mr. James Taylors rowling road, 91

road from Terrys run up to the mountain road ...above Mr Richard Bayleys land and houses, 60, 61, 64

Road from Thorntons Quarter to Mitchels ford upon the river by Collo: Henry Willis Quarter up to the upper inhabitants in the little fork, 63

Road from Todds branch to the Robinson, 73

road from John Wallers bridge to Mattapony Church, 24, 59, 82

bridle Way to go over the ford by mr: Walkers Quarter over the River Po to Church, 50

road from the Walnut Branch on ye. North Side of Rappadan Down the Ridge & to Cross ye. Rappadan so Down to the South West Mountain Chappell, 39, 44, 50

road from the Wilderness in the lower precincts of the old mountain road, 63

road from the Wilderness bridge to Germanna, 18, 76, 77

road from the Wilderness Bridge to the Pine Stake, 8, 23, 90, 92

road from Henry Willis's mill to the Courthouse over muddey run, 37

road from Collo: Henry Willis's Mill in ye. fork of Rappahannock River to Germanna, 43^2, 47^3, 52, 53

road from Henry Willis's plantation called Byram's plantation to the main road at Fredericksburgh, 63, 66

road from Colo. Willis's Upper Quarter to Battle Mountain, 76

road from Benjamin Winslows to the first Main road, 77

Road from Woodfords Quarter to Mr Francis Thorntons house, 18

www.ingramcontent.com/pod-product-compliance
Lightning Source LLC
Chambersburg PA
CBHW080250170426
43192CB00014BA/2629